MULTIPLE CHOICE AND

Higher **BIOLOGY**

for CfE

Writing Team:

James Torrance

James Fullarton

Clare Marsh

James Simms

Caroline Stevenson

Diagrams by James Torrance

SCOTTISH
EXAMINATION
MATERIALS

HODDER
GIBSON
AN HACHETTE UK COMPANY

The Publishers would like to thank the following for permission to reproduce copyright material:

Photo credits

p.1 (background) and Unit 1 running head image © 2010 Steve Allen/Brand X Pictures/photolibrary.com; p.1 (inset left) © april21st – Fotolia, (inset centre) © Dr Keith Wheeler/Science Photo Library, (inset right) James Torrance; p.49 (background) and Unit 2 running head image © Loren Rodgers – Fotolia; p.49 (inset left) © Natural Visions/Alamy, (inset centre) © Dennis Kunkel Microscopy, Inc/Visuals Unlimited, Inc., (inset right) © Dennis Kunkel Microscopy, Inc./Visuals Unlimited/Corbis; p.105 (background) and Unit 3 running head image © pro6x7 – Fotolia; p.105 (inset left) © Anna – Fotolia, (inset centre) © Jef Meul/Minden Pictures/FLPA RM, (inset right) © Oceans-Image/Photoshot.

Every effort has been made to trace all copyright holders, but if any have been inadvertently overlooked the Publishers will be pleased to make the necessary arrangements at the first opportunity.

Hachette UK's policy is to use papers that are natural, renewable and recyclable products and made from wood grown in sustainable forests. The logging and manufacturing processes are expected to conform to the environmental regulations of the country of origin.

Orders: please contact Bookpoint Ltd, 130 Park Drive, Milton Park, Abingdon, Oxon OX14 4SE. Telephone: (44) 01235 827720. Fax: (44) 01235 400454. Lines are open 9.00–5.00, Monday to Saturday, with a 24-hour message answering service. Visit our website at www.hoddereducation.co.uk. Hodder Gibson can be contacted direct on: Tel: 0141 848 1609; Fax: 0141 889 6315; email: hoddergibson@hodder.co.uk

© James Torrance, James Fullarton, Clare Marsh, James Simms, Caroline Stevenson 2015

First published in 2015 by

Hodder Gibson, an imprint of Hodder Education,

An Hachette UK Company,

2a Christie Street

Paisley PA1 1NB

Impression number   5 4 3 2 1

Year   2019 2018 2017 2016 2015

Cover photo © Andy Rouse/naturepl.com

Illustrations by James Torrance

Typeset in Minion Regular 11/14pt by Integra Software Services Pvt. Ltd., Pondicherry, India

Printed in Slovenia

A catalogue record for this title is available from the British Library

ISBN: 978 1 4718 4742 4

Contents

Unit 1

DNA and the Genome

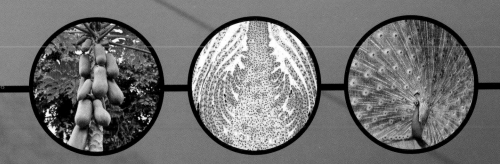

1 Structure of DNA

Matching test

Match the terms in list X with their descriptions in list Y.

list X
1 3' end *e*
2 5' end *d*
3 adenosine (A) *m*
4 antiparallel *g*
5 cytosine (C) *j*
6 deoxyribose *i*
7 DNA *o*
8 double helix *p*
9 eukaryote *a*
10 genotype
11 guanine (G) *l*
12 nucleotide *f*
13 plasmid
14 prokaryote *b*
15 sugar–phosphate backbone *h*
16 thymine (T) *k*

list Y
a) advanced organism such as a plant or animal
b) simple organism such as a bacterium
c) small ring of DNA present in prokaryotes
d) phosphate end of DNA strand to which nucleotides cannot be added
e) deoxyribose end of DNA strand to which nucleotides can be added
f) basic unit of which nucleic acids are composed
g) the relationship between two strands of DNA with their sugar-phosphate backbones running in opposite directions
h) supporting structure of nucleic acid molecule formed by bonding between adjacent nucleotides
i) sugar present in DNA
j) base present in DNA which is complementary to guanine
k) base present in DNA which is complementary to adenine
l) base present in DNA which is complementary to cytosine
m) base present in DNA which is complementary to thymine
n) nucleic acid present in linear and circular chromosomes
o) genetic constitution of an organism determined by the sequence of bases in its DNA
p) two-stranded molecule of DNA wound into a spiral

Multiple choice test

Choose the ONE correct answer to each of the following multiple choice questions.

1   The structure of one nucleotide is shown below.

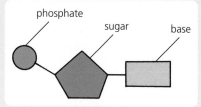

Figure 1.1

Which of the following diagrams shows two nucleotides correctly joined together?

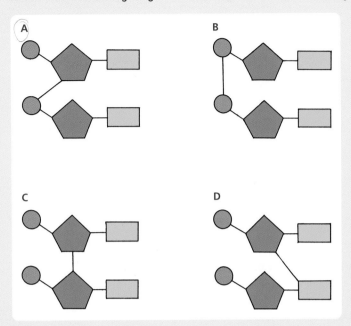

Figure 1.2

2 If a double-stranded DNA molecule is 50 000 base pairs long, how many nucleotides does it contain?

A 25 000   B 50 000   C 100 000   D 200 000

3 If a DNA molecule contains 10 000 base molecules of which 18% are thymine, then the number of cytosine molecules present is

A 1800   B 3200   C 6400   D 8200

4 If a DNA molecule contains 4000 base molecules and 1200 of these are adenine, then the percentage number of guanine bases present in the molecule is

A 12   B 20   C 28   D 30

5 The following diagram shows part of one strand of a DNA molecule.

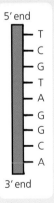

5′ end
T
C
G
T
A
G
G
C
A
3′ end

Figure 1.3

Which of the strands in the diagram below is the complement of the original strand?

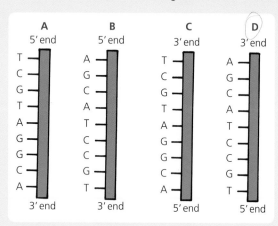

Figure 1.4

6   The diagram below shows the stages that occur in an actively dividing mammalian cell.

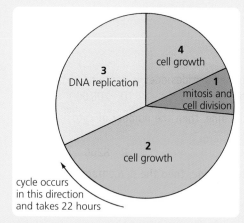

Figure 1.5

The drug aminopterin inhibits thymine production. When this drug is added to a culture of actively dividing cells, after 16 hours most of the cells are found to have been UNABLE to complete one of the stages in the cycle. This stage is number

A   1   B   2   C   3   D   4

7 The diagram below represents Griffith's bacterial transformation experiment.

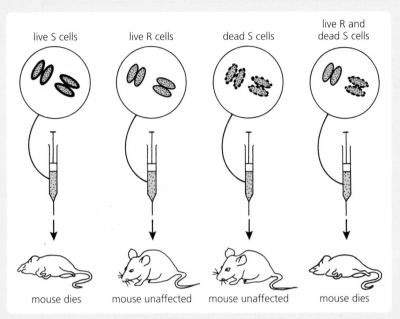

Figure 1.6

Which syringe in the following diagram would NOT lead to the death of a mouse?

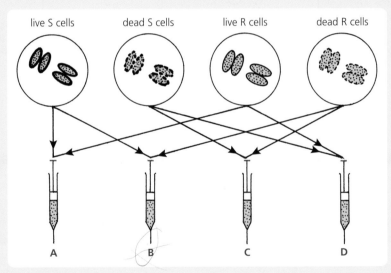

Figure 1.7

8 Applying Chargaff's rules relating to the number of different bases in a DNA sample, which of the following is correct?
   A T = C   B G = A   C C + G = T + A   D T + C = G + A

9 A bacterial cell is
   A eukaryotic and all of its DNA is circular.   B prokaryotic and all of its DNA is circular.
   C eukaryotic and all of its DNA is linear.   D prokaryotic and all of its DNA is linear.

10 The accompanying diagram shows four stages employed during the separation of DNA by gel electrophoresis.
   The order in which they would be carried out is
   A 4,2,1,3   B 2,4,3,1   C 2,4,1,3   D 4,2,3,1

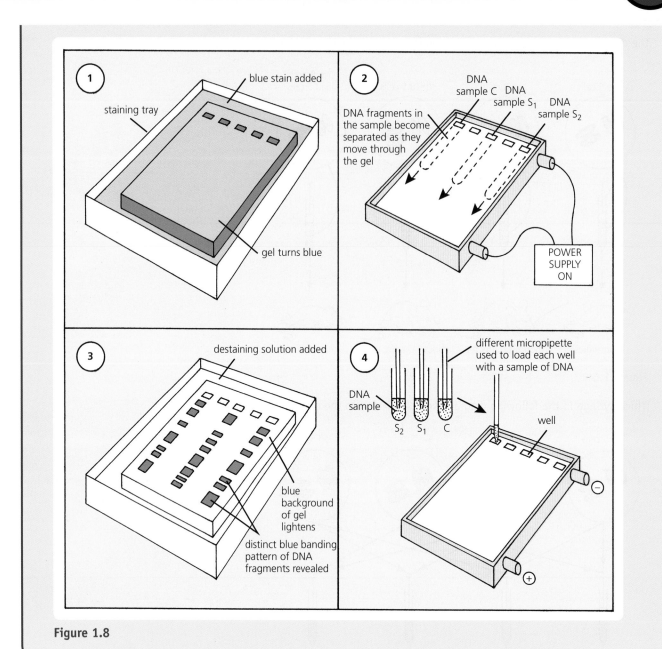

1 blue stain added
staining tray
gel turns blue

2 DNA sample C
DNA sample S₁
DNA sample S₂
DNA fragments in the sample become separated as they move through the gel
POWER SUPPLY ON

3 destaining solution added
blue background of gel lightens
distinct blue banding pattern of DNA fragments revealed

4 different micropipette used to load each well with a sample of DNA
DNA sample
S₂ S₁ C
well

Figure 1.8

2 Replication of DNA

Matching test
Match the terms in list X with their descriptions in list Y.

list X
1 amplification
2 complementary
3 DNA
4 DNA polymerase
5 ligase
6 polymerase chain reaction (PCR)
7 primer
8 replication
9 specific target sequence
10 template strand

list Y
a) complex molecule present in chromosomes which stores genetic information
b) term used to refer to the unwound strand of DNA to be replicated
c) the relationship between two members of a base pair able to join by hydrogen bonding
d) increase in number of copies of a DNA molecule by PCR
e) process by which a molecule of DNA reproduces itself
f) enzyme required to promote DNA replication
g) laboratory technique used to create many copies of a piece of DNA
h) enzyme that joins replicated DNA fragments into a complete strand
i) short sequence of nucleotides needed by DNA polymerase to begin replication of DNA
j) region at the end of a DNA strand complementary to a primer

Multiple choice test
Choose the ONE correct answer to each of the following multiple choice questions.

Questions 1, 2, and 3 refer to the following diagram which shows the formation of the lagging strand of DNA during replication.

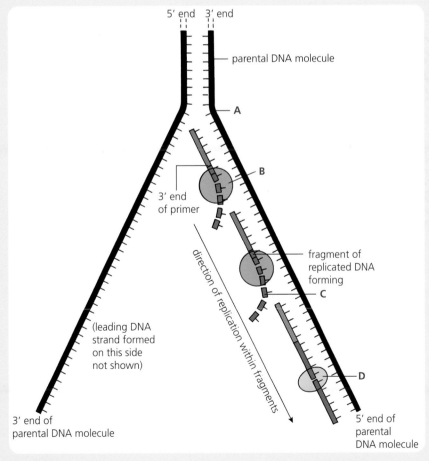

5' end   3' end

parental DNA molecule

A

B

3' end
of primer

fragment of
replicated DNA
forming

C

direction of replication within fragments

(leading DNA
strand formed
on this side
not shown)

D

3' end of
parental DNA molecule

5' end of
parental
DNA molecule

Figure 2.1

1 Which structure is ligase? *D*
2 Which structure is DNA polymerase? *B*
3 Which structure is part of a replication fork? *A*
4 The following diagram shows a portion of DNA undergoing semi-conservative replication.

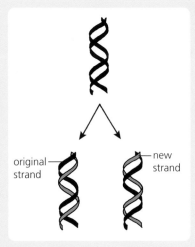

original strand

new strand

Figure 2.2

If the products undergo a further round of semi-conservative replication, which of the sets of products shown in the diagram below would be the result?

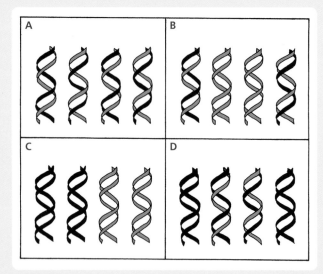

Figure 2.3

5 The steps given in the following list are involved in the first cycle of the PCR technique.
   1 Heat-tolerant DNA polymerase adds nucleotides to the primers.
   2 DNA to be amplified is heated to separate its two strands.
   3 Two identical copies of the original DNA molecule are formed.
   4 Following cooling, each primer binds to its target DNA sequence.
   Their correct sequence is

   A 1,4,3,2   B 2,1,3,4   C 2,4,1,3   D 4,2,3,1

6 The following diagram represents the early stages of the first cycle of PCR.

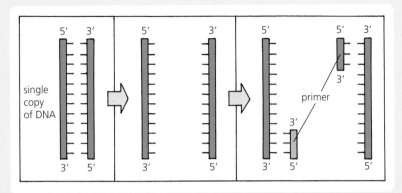

Figure 2.4

Which of the answers in the diagram below shows the correct outcome at the end of the first cycle?

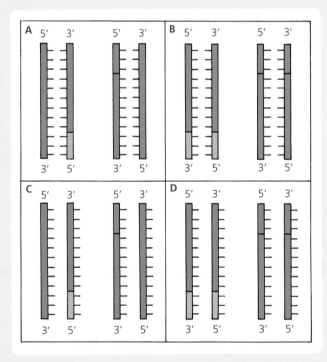

Figure 2.5

7 Starting with a single DNA molecule, how many DNA molecules would be present after 8 cycles of the PCR procedure?

   A 16   **B** 64   **C** 256   **D** 512

Questions 8 and 9 refer to the information in Figure 2.6 which shows three types of DNA molecule. ——— represents a single DNA strand labelled with ^{15}N (a heavy isotope of nitrogen) and - - - - - represents a single DNA strand labelled with ^{14}N (a common isotope of nitrogen). A mixture of these can be separated as indicated.

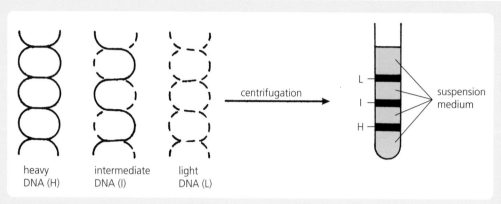

Figure 2.6

8 If bacteria, grown for many generations in ^{15}N, are transferred to ^{14}N medium for one generation's growth, which of the following will result on extracting and centrifuging their DNA?

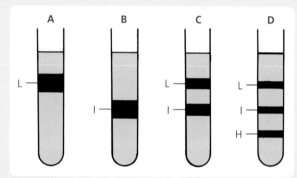

Figure 2.7

9 Imagine that a molecule of heavy DNA is replicated for two successive generations in ^{14}N medium. Which line in the following table shows the DNA molecules that would be formed?

	heavy DNA	intermediate DNA	light DNA
A	0	100%	0
B	50%	0	50%
C	0	50%	50%
D	50%	50%	0

Table 2.1

10 Evolutionary relationships between different groups of plants can be investigated following amplification of their DNA by

A gel electrophoresis.

C degradation by protease.

B precipitation in alcohol.

D polymerase chain reaction.

3 Control of gene expression

Matching test part 1
Match the terms in list X with their descriptions in list Y.

list X
1 amino acid
2 antibody
3 chromatography
4 enzyme
5 gel electrophoresis
6 genotype
7 hormone
8 hydrogen bond
9 nitrogen
10 peptide bond
11 phenotype
12 polypeptide
13 protein

list Y
a) chemical element present in all proteins in addition to carbon, hydrogen and oxygen
b) weak chemical link holding a polypeptide chain in a coil
c) strong chemical link joining adjacent amino acids in a polypeptide chain
d) type of protein that acts as a chemical messenger
e) type of protein made by white blood cells to defend the body against antigens
f) type of protein possessing an active surface which combines with a specific substrate
g) chain-like molecule composed of several amino acids
h) molecule composed of one or more polypeptides folded or coiled into a specific shape
i) physical and chemical state of a cell or organism produced as a result of gene expression
j) genetic constitution of a cell or organism determined by the sequence of bases in its DNA
k) one of twenty different types of organic compound which are the basic building blocks of protein
l) technique used to separate components of a mixture which differ in their degree of solubility in a solvent
m) technique used to separate electrically charged molecules by subjecting them to an electric current which forces them to move through a sheet of gel

Matching test part 2
Match the terms in list X with their descriptions in list Y.

list X
1 anticodon
2 attachment site
3 codon
4 exon
5 genetic code
6 intron
7 mRNA
8 post-translational modification
9 primary transcript
10 ribose
11 ribosome
12 RNA polymerase
13 splicing
14 transcription
15 translation
16 tRNA
17 uracil

list Y
a) molecular language made up of 64 codewords
b) conversion of the genetic information on mRNA into a sequence of amino acids in a polypeptide
c) process by which a complementary molecule of mRNA is made from a region of a DNA template
d) type of nucleic acid which carries a specific amino acid to a ribosome
e) type of nucleic acid which carries a copy of the DNA code from the nucleus to a ribosome
f) base present in RNA that is complementary to adenine
g) sub-cellular structure made of rRNA and protein which is the site of protein synthesis
h) sugar present in RNA
i) enzyme which controls transcription
j) triplet of bases on a tRNA molecule which is complementary to an mRNA codon
k) unit of genetic information consisting of three mRNA bases
l) coding region of a gene
m) non-coding region of a gene
n) mRNA strand formed as the complement of a DNA template strand
o) region on a tRNA molecule to which a specific amino acid becomes temporarily fixed
p) joining of exons from a primary transcript of RNA following the removal of introns
q) alteration of protein molecule by cutting and combining its polypeptides

➡

Multiple choice test
Choose the ONE correct answer to each of the following multiple choice questions.

1 The number of different types of amino acid commonly found to make up proteins is approximately

   A 20　　　　　　　　　　B 64　　　　　　　　　　C 200　　　　　　　　　　D 640

2 The connections that join individual amino acid molecules into a long chain are called

   A hydrogen bonds.　　　　　　　　　　　　　B protein bonds.
   C peptide bonds.　　　　　　　　　　　　　　D sugar-phosphate bonds.

Questions 3 and 4 refer to the accompanying diagram which shows a chromatogram of five amino acids. The Rf value of an amino acid is the ratio of the distance moved by the amino acid front from the origin to the distance moved by the solvent front from the origin.

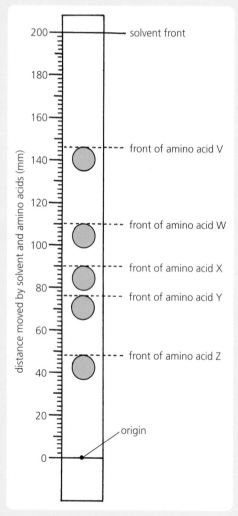

Figure 3.1

3 Which amino acid has an Rf value of 0.55?

   A V　　　　　　　　　　B W　　　　　　　　　　C X　　　　　　　　　　D Z

4 The Rf value of amino acid Y is

   A 0.34　　　　　　　　　B 0.38　　　　　　　　　C 0.68　　　　　　　　　D 0.76

Questions 5, 6 and 7 refer to the following information. During gel electrophoresis, electrically charged molecules are subjected to an electric current that forces them to move through a sheet of gel. The smaller the molecule, the further the distance that it travels. Figure 3.2 shows the result of employing this technique to produce protein 'fingerprints' of five samples taken from fish P, Q, R, S and T.

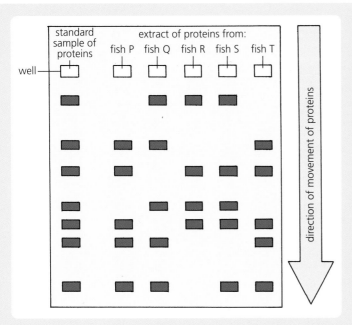

Figure 3.2

5 Which of the following fish protein extracts differs in THREE ways from the standard sample of fish proteins?

  A P   **B** Q   **C** R   **D** S

6 Which of the following pairs of fish protein extracts differ from one another in FOUR ways?

  A P and R   **B** P and S   **C** Q and R   **D** R and T

7 Which pair of protein extracts is most likely to belong to the same species of fish?

  A P and T   **B** R and S   **C** Q and R   **D** P and Q

8 The following table gives the mass per 100g of protein of five different amino acids found in four proteins.

	protein	mass of amino acid (g/100g protein)				
		glycine	alanine	leucine	valine	phenylalanine
A	insulin	4	5	13	8	8
B	haemoglobin	6	7	15	9	8
C	keratin	7	4	11	5	4
D	albumin	3	7	9	7	8

Table 3.1

Which protein is represented by the following pie chart?

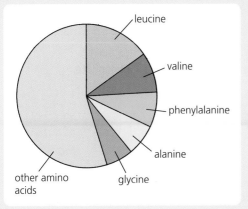

Figure 3.3

9 The accompanying diagram shows the sequence of amino acids present in one molecule of insulin. In this protein the ratio of leucine: glycine: tyrosine: histidine is

A 6: 4: 3: 2   B 6: 4: 4: 1   C 3: 2: 1: 1   D 3: 2: 2: 1

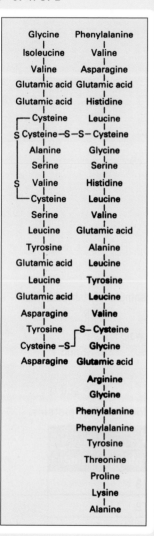

Figure 3.4

10 Choose the one correct pair of answers needed to complete the sentence.

Nucleotides are to _____ as _____ are to proteins.

A ribosomes; amino acids   B nucleic acids; enzymes

C ribosomes; enzymes   D nucleic acids; amino acids

11 One of the nucleotides present in mRNA has the composition

A adenine – ribose – phosphate.   B uracil – deoxyribose – phosphate.

C thymine – ribose – phosphate.   D guanine – deoxyribose – phosphate.

12 Which line in the following table is correct?

	present in DNA	present in RNA
A	uracil	thymine
B	ribose	deoxyribose
C	double strand	single strand
D	four different nucleotides	five different nucleotides

Table 3.2

Questions 13 and 14 refer to the following diagram which represents transcription of a small portion of mRNA from a template strand of DNA.

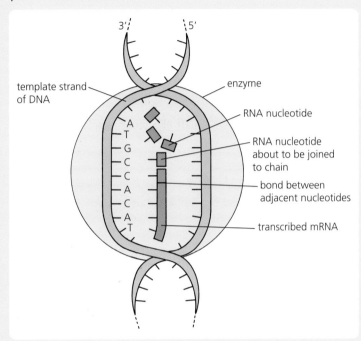

Figure 3.5

13 The enzyme responsible for this process is

   A ligase.   **B** polypeptidase.   **C** DNA polymerase.   **D** RNA polymerase.

14 Which part of the following diagram correctly shows the transcribed mRNA?

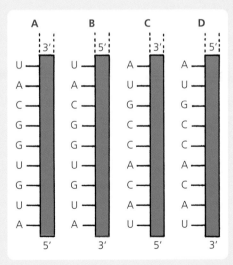

Figure 3.6

15 A free transfer RNA molecule can combine with

   A one specific amino acid only.   **B** any available amino acid.

   C three different amino acids.   **D** a chain of amino acids.

Questions 16, 17, 18 and 19 refer to the following possible answers.

   A DNA   **B** tRNA   **C** mRNA   **D** amino acid

16 On which type of molecule are anticodons found?

17 Which of these molecules bears codons which are complementary to anticodons?

18 Which of these molecular types must be present in the largest number for successful synthesis of a large protein molecule to occur?

19 Which of these types of molecule holds the master copy of the genetic information which determines the order in which amino acids are joined into a growing protein chain?

Questions 20 and 21 refer to the following diagram which shows the synthesis of part of a protein molecule.

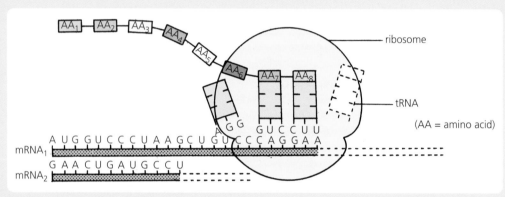

mRNA₁   A U G G U C C C U A A G C U G U C C C A G G A A

mRNA₂   G A A C U G A U G C C U

ribosome

tRNA

(AA = amino acid)

Figure 3.7

20 Which of the following is the first part of the protein molecule that would be translated from mRNA₂?

   A AA₄ – AA₂ – AA₇ – AA₆   **B** AA₆ – AA₇ – AA₂ – AA₄

   C AA₃ – AA₁ – AA₅ – AA₈   **D** AA₈ – AA₅ – AA₁ – AA₃

21 The following diagram shows a small part of a different protein that was also synthesised on this ribosome.

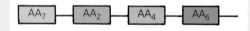

Figure 3.8

What sequence of bases in DNA coded for this sequence of amino acids?

   A CAGGUCAAGUCC   **B** GTCCAGTTCAGG

   C GUCCAGUUCAGG   **D** CAGCTCAAGTCC

22 If each amino acid weighs 100 mass units, what is the weight (in mass units) of the protein molecule synthesised from an mRNA molecule which is 600 bases long?

   A 2000   **B** 6000   **C** 20 000   **D** 60 000

23 A variety of different proteins can be expressed from the same gene as a result of

   A alternative splicing of RNA and pretranslational modification.

   B alternative splicing of DNA and pretranslational modification.

   C alternative splicing of RNA and post-translational modification.

   D alternative splicing of DNA and post-translational modification.

24 The graph below gives the results of an investigation into the relative number of ribosomes present in the cells of a new leaf developing at a shoot tip.

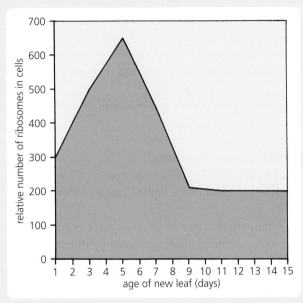

Figure 3.9

Which of the following is the LEAST likely explanation of the results?

A The period of most rapid protein synthesis occurred between days 1 and 3.

B After day 5 protein synthesis and growth slowed down.

C Growth of this leaf was complete by day 11.

D Protein synthesis had come to a halt by day 15.

25 When a cell's DNA becomes damaged, phosphate is added to a regulatory protein called p53 making it become an active p53 tumour-suppressor protein. This process is an example of

A cleavage of exons.

B alternative RNA splicing.

C transcription of introns.

D post-translational modification.

4 Cellular differentiation

Matching test
Match the terms in list X with their descriptions in list Y.

list X
1 blastocyst
2 bone marrow
3 differentiation
4 embryonic stem cell
5 ethics
6 meristem
7 regulation
8 specialised
9 therapeutic
10 tissue (adult) stem cell
11 tissue culture

list Y
a) form of control to ensure the quality of stem cells used and the safety of the procedures carried out
b) type of artificial propagation used to produce a clone of a desirable plant
c) moral values that ought to govern human conduct
d) common source of tissue (adult) stem cells in humans
e) early human embryo composed of unspecialised cells
f) term describing the use of stem cells in the repair of damaged or diseased organs
g) type of stem cell capable of differentiating into a limited range of an animal's cell types
h) type of stem cell capable of differentiating into all of an animal's cell types
i) group of unspecialised plant cells capable of dividing repeatedly throughout the life of the plant
j) process of cell specialisation involving the selective switching off and on of certain genes
k) cell that has become differentiated and only expresses genes for proteins specific to that cell type

Multiple choice test
Choose the ONE correct answer to each of the following multiple choice questions.

1 Meristematic cells are
   A found only at root and shoot tips in plants.
   B unspecialised and capable of dividing repeatedly.
   C widely distributed throughout a developing animal's body.
   D used to transport materials in the stem of a green plant.

2 In a horse chestnut tree, the cells of a newly opened bud do not give rise to a stem or root because such cells
   A only possess the genes required to form leaves.
   B are located at an elevated position on the plant.
   C have only the genes for leaf formation switched on.
   D suffer loss of root and stem genes during cell division.

Questions 3, 4 and 5 refer to Figure 4.1 which shows an experiment carried out to investigate the effect of a growth regulator on root cells from a type of rare tree.

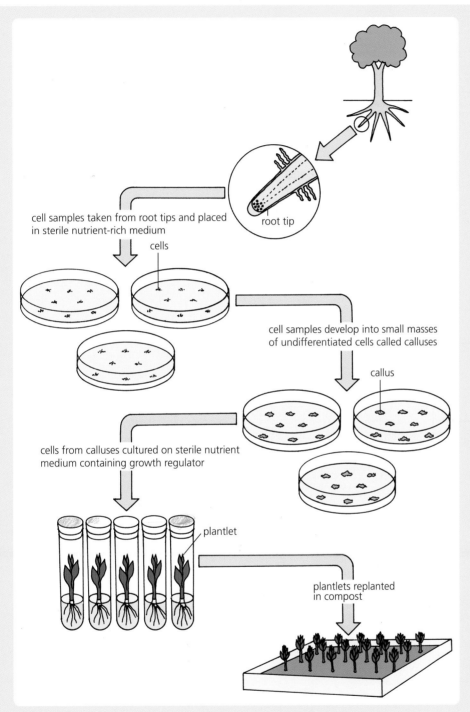

Figure 4.1

3 Which of the following occurs during the course of events shown in the diagram?
   A tissue-culturing a plant clone
   C sterilisation of stem cells
   B differentiation of embryonic cells
   D transfer of nuclei between two species

4 Which of the following would be the valid control for this experiment?
   A shoot tip cells in nutrient-rich medium
   B root tip cells in medium lacking nutrients
   C cells from calluses in medium lacking growth regulator
   D plantlets replanted in sterilised potting compost

5   The inclusion of replicates in this experiment enables the investigator to
   A   increase the reliability of the results.
   B   include an effective set of controls.
   C   investigate several plant species at the one time.
   D   study the effects of several variable factors simultaneously.

6   Each cell in a multicellular organism contains all the genes necessary for the construction of
   A   that one cell only.   B   all cells of that type of tissue.
   C   all types of that cell.   D   the whole organism.

Questions 7 and 8 refer to the following diagram which represents the processes of cell division and cellular differentiation in an animal.

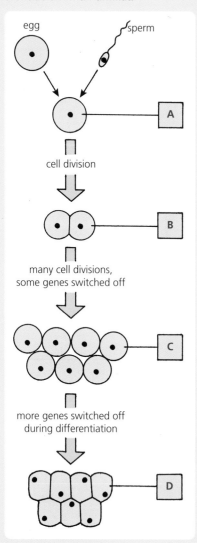

Figure 4.2

7   Which box indicates a tissue (adult) stem cell?
8   Which box indicates an embryonic stem cell?
9   Which of the following statements about stem cells is NOT true?
   A   They are unspecialised cells present in multicellular animals.
   B   They are able to reproduce themselves by repeated mitosis and cell division.
   C   They differentiate into specialised cells when required to do so.
   D   They are meristematic cells present in the stems of flowering plants.

10 The five boxes in the following diagram describe stages in the procedure carried out during bone marrow transplantation.

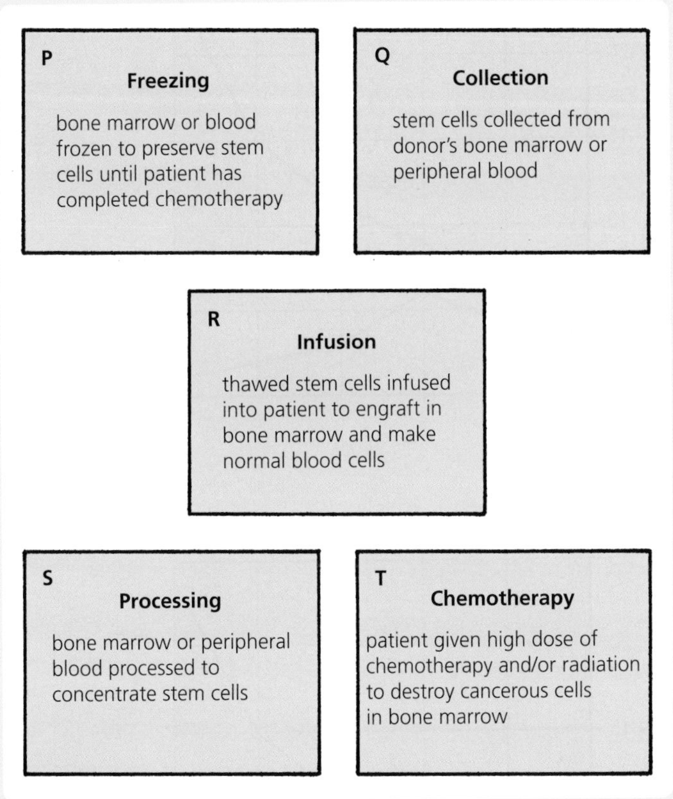

| P | Freezing |
bone marrow or blood frozen to preserve stem cells until patient has completed chemotherapy

| Q | Collection |
stem cells collected from donor's bone marrow or peripheral blood

| R | Infusion |
thawed stem cells infused into patient to engraft in bone marrow and make normal blood cells

| S | Processing |
bone marrow or peripheral blood processed to concentrate stem cells

| T | Chemotherapy |
patient given high dose of chemotherapy and/or radiation to destroy cancerous cells in bone marrow

Figure 4.3

Their correct sequence is

A S, Q, T, R, P   B Q, S, P, T, R   C S, Q, P, R, T   D Q, P, S, T, R

11 Under natural conditions, a tissue (adult) stem cell of a human being has
A all of its genes switched on and is capable of differentiation into any cell type.
B all of its genes switched on but is only capable of giving rise to a limited range of cell types.
C many of its genes switched off but is capable of differentiating into any cell type.
D many of its genes switched off and is only capable of giving rise to a limited range of cell types.

12 The table shows the result of an experiment set up to investigate the effect of a growth-stimulating factor on a type of stem cell. Flask 1 contained nutrient-rich medium + growth stimulating factor. Flask 2 contained nutrient-rich medium only. Which of the graphs in Figure 4.4 represents the results correctly?

time (days)	mean cell number x 10^5/cm³	
	flask 1	flask 2
0	3	3
2	4	5
4	9	6
6	11	3
8	15	2
10	24	1

Table 4.1

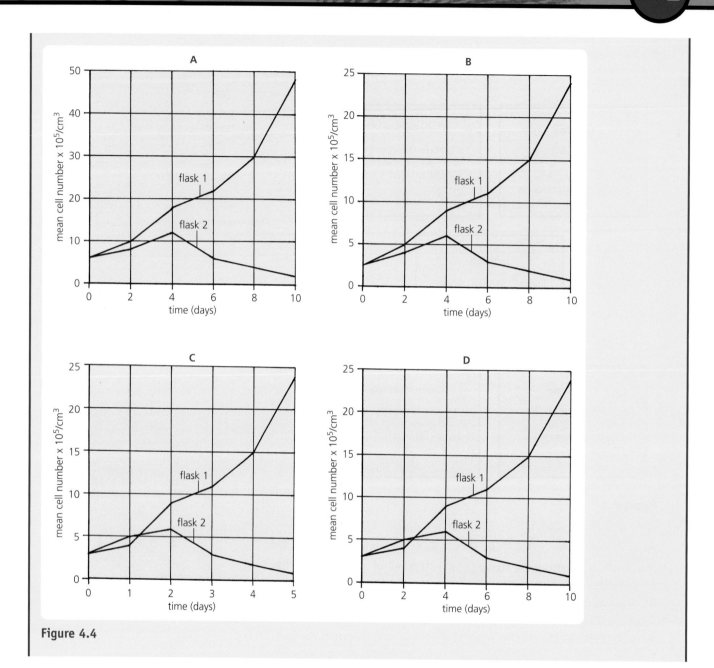

Figure 4.4

Questions 13 and 14 refer to the following diagram which shows a simplified version of the means by which 'Dolly, the sheep' was produced.

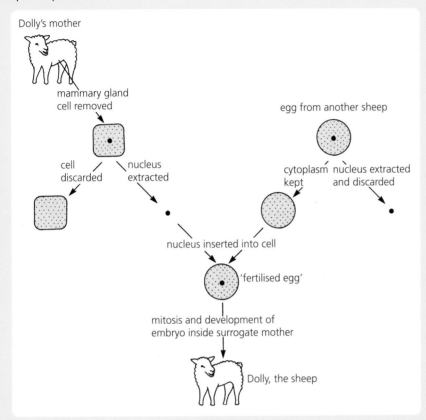

Figure 4.5

13 The name of the process employed to create the original cell that gave rise to Dolly is called
  A meristematic cell-culturing.
  B nuclear transfer technique.
  C transplantation of embryonic cell.
  D induction of pluripotent stem cell.

14 Which of the following statements is FALSE?
  A Dolly and her mother contained genetically identical nuclear material.
  B A mammary gland cell in a sheep contains repressed genes that can become switched on again.
  C Differentiation of the mammary gland cells in a sheep is an irreversible process.
  D During Dolly's embryonic development, cells became differentiated and specialised.

15 The use of stem cells raises several ethical issues.
  Ethics refers to
  A the moral values that ought to govern human conduct.
  B the techniques followed when a procedure is closely regulated.
  C the safety standards that must be maintained during research work.
  D the qualifications of the experts employed to carry out an investigation.

5 Structure of the genome & 6 Mutations

Matching test part 1

Match the terms in list X with their descriptions in list Y.

list X	list Y
1 deletion	a) change affecting nucleotide(s) at a site where introns are normally removed from a primary mRNA transcript
2 gene	
3 genome	b) physical and biochemical characteristics of an organism resulting from the expression of its genes
4 insertion	
5 mutant	c) term referring to a change in one of the base pairs in the DNA sequence of a single gene
6 mutation	d) gene mutation involving the loss of one nucleotide from the DNA sequence
7 non-coding sequence	e) gene mutation involving the addition of an extra nucleotide to the DNA sequence
8 phenotype	f) non-coding sequence of DNA bearing activators needed to promote transcription of a gene
9 point mutation	g) individual whose genotype expresses a mutation
10 regulator sequence	h) gene mutation involving the exchange of one nucleotide for another in the DNA sequence
11 splice-site mutation	i) sequence of DNA bases that codes for a protein
12 substitution	j) general name for a region of DNA that does not code for protein
	k) general term for a random change in an organism's genome
	l) an organism's entire genetic information encoded in the DNA of a complete set of its chromosomes

Matching test part 2

Match the terms in list X with their descriptions in list Y.

list X	list Y
1 chromosome structure mutation	a) protective region of repeated DNA sequences found at the end of a chromosome
	b) doubling up of part of a chromosome involving several genes
2 deletion	c) general term referring to a change that alters the number or sequence of the genes in a chromosome
3 duplication	
4 inversion	d) transfer of a segment of genes from one chromosome to another non-matching chromosome
5 polyploidy	
6 telomere	e) loss of a segment of chromosome consisting of one or more genes
7 translocation	f) reversal of the gene order of a segment of chromosome as a result of two breaks in the same chromosome
	g) form of mutation involving one or more extra sets of chromosomes being added to a species' chromosome complement

Multiple choice test

Choose the ONE correct answer to each of the following multiple choice questions.

1 The accompanying diagram shows part of a DNA strand containing a gene to be transcribed.

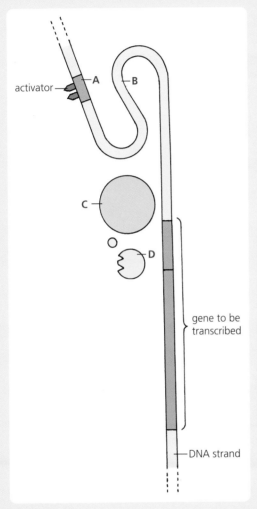

activator —

gene to be transcribed

DNA strand

Figure 5.1

Which structure is the DNA sequence that will regulate this process?

Questions 2, 3 and 4 refer to the following possible answers.

   A tRNA   B mRNA   C rRNA   D miRNA

2 Which form of RNA combines with protein molecules to form the complexes of which ribosomes are composed?

3 Which form of RNA undergoes folding into a clover leaf-shaped molecule to which a molecule of amino acid becomes attached?

4 Which form of RNA brings about the silencing of a gene following the gene's transcription?

5 A mutation is a

   A change in genotype which may result in a new expression of a characteristic.

   B sudden temporary change in an organism's genetic material.

   C change in phenotype followed by a change in genotype.

   D change in hereditary material directed by a changing environment.

6 Which of the following statements is NOT correct?
  A Mutations provide variation upon which natural selection can act.
  B Mutation rate can be increased by artificial means.
  C Mutations arise spontaneously, infrequently and at random.
  D The vast majority of mutations produce alleles which are dominant.

7 The table shows the results from an investigation into the effect of increased dosage of X-rays on the frequency of lethal mutations in fruit flies.

dosage of X-rays (roentgens)	frequency of lethal mutations (%)
1000	3
2000	5
4000	13
8000	23

Table 5.1

Which of the following graphs correctly represents the results as a line of best fit?

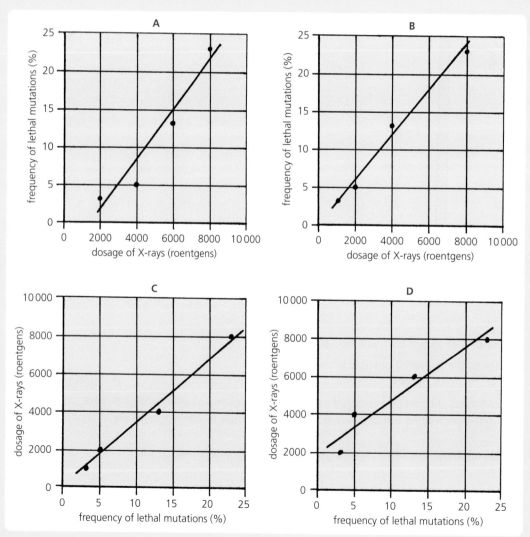

Figure 5.2

8 Which of the following is NOT a mutagenic agent?
  A mustard gas   B gamma rays   C low temperature   D ultraviolet rays

9 The following diagram shows regulation of a normal unaltered protein-coding gene.

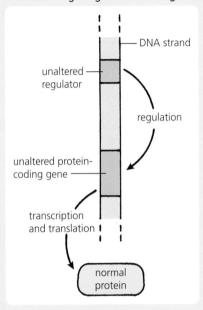

Figure 5.3

Which arrow in the diagram below represents the process of regulation that could result in the transcription and translation of a non-functional version of the protein?

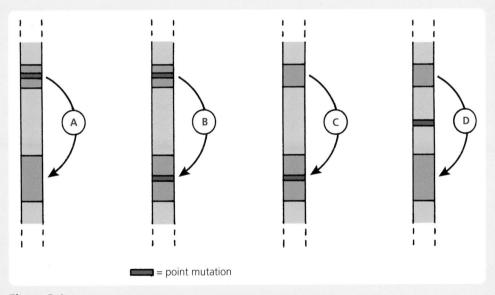

Figure 5.4

Questions 10, 11 and 12 refer to the following possible answers.

   A inversion   **B** deletion   **C** insertion   **D** substitution   ➡

10 What name is given to the type of point mutation illustrated in the following diagram?

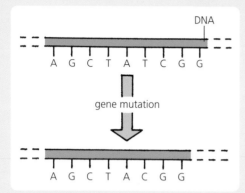

Figure 5.5

11 What name is given to the type of point mutation shown in the following diagram?

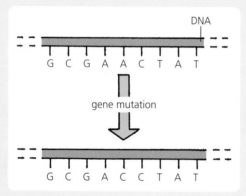

Figure 5.6

12 What name is given to the type of point mutation illustrated in the following diagram?

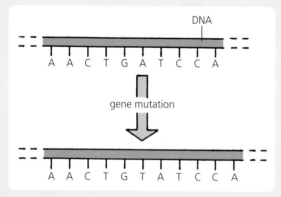

Figure 5.7

13 Which of the following point mutations BOTH lead to a major change that causes a large portion of the gene's DNA to be misread?

  A deletion and insertion   B insertion and inversion

  C inversion and substitution   D substitution and deletion

14 Neurofibromatosis, a condition in which the human sufferer develops multiple brown lumps in the skin, is caused by a dominant mutant allele whose mutation rate is 100 per million gametes.

The chance of a new mutation occurring is therefore

  A 1 in 1000   B 1 in 10 000   C 1 in 100 000   D 1 in 1 000 000

15 The amino acid sequences shown in the following diagram belong to a particular polypeptide chain present in the wild type variety and several mutant strains of a species of bacterium.
Which mutant strain has undergone an insertion of a nucleotide into its DNA sequence?

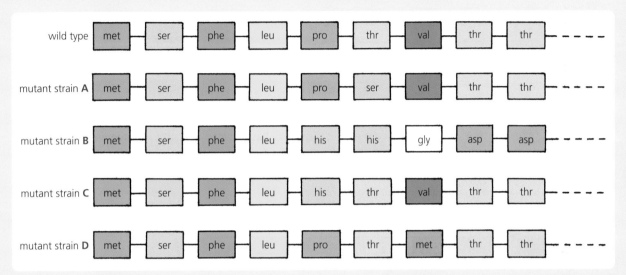

Figure 5.8

16 The following diagram shows the outcome of a cross between two sufferers of sickle cell trait (where H = the allele for normal haemoglobin and S = the allele for haemoglobin S).

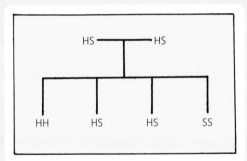

Figure 5.9

With respect to survival of the offspring, which line in the table below would be MOST likely?

	% number of survivors	
	population living in malarial area	population living in non-malarial area
A	25	75
B	50	75
C	25	100
D	50	100

Table 5.2

17 The diagram below shows a pair of matching chromosomes during gamete formation.

Figure 5.10

Which of the following terms refers to the type of mutation that has affected the altered chromosome?

A deletion **B** inversion **C** duplication **D** translocation

18 The following diagram shows the outcome of a certain type of chromosome mutation. The lettered regions indicate the positions of six marker genes.

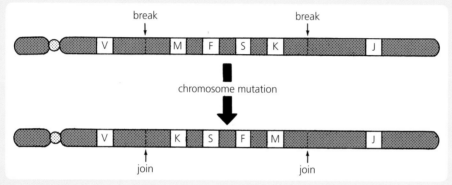

Figure 5.11

Which of the diagrams below best represents this mutated chromosome paired with its unaltered matching partner during gamete formation?

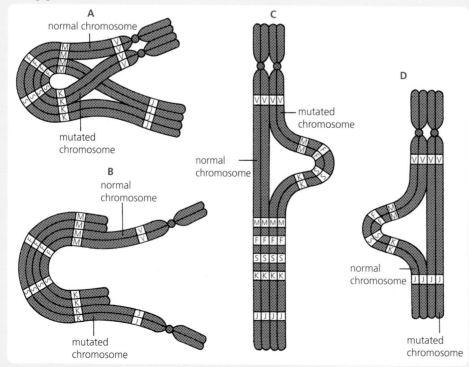

Figure 5.12

Questions 19 and 20 refer to the following diagram where chromosomes 1 and 2 are undergoing mutations.

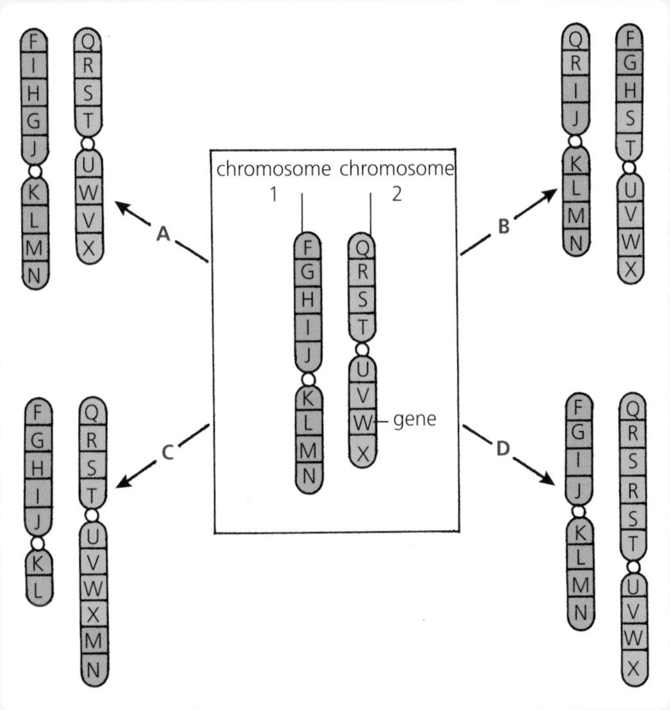

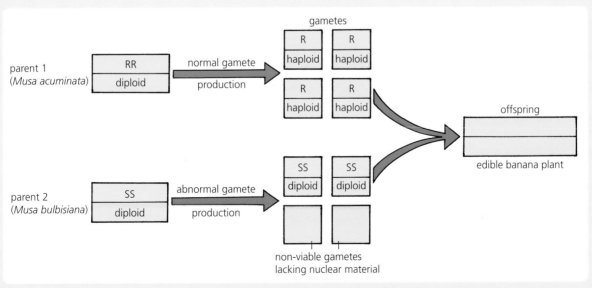

Figure 5.13

19 Which arrow represents reciprocal translocation?

20 Which arrow represents non-reciprocal translocation?

21 A species of plant is known to have a diploid chromosome number of 14 in each of its cells. Which of the following would be the number of chromosomes found in each cell of one of its polyploid relatives?

   A 7   **B** 14   **C** 28   **D** 47

22 Polyploid wheat does NOT normally show an increase in

   A size.   **B** vigour.

   C resistance to disease.   **D** length of life cycle.

23 The following diagram shows how a variety of edible banana arose.

Figure 5.14

Which of the boxes in the diagram below contains the correct information needed to complete the first diagram?

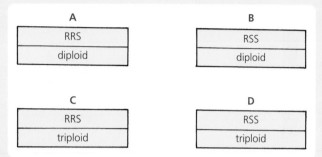

A	B
RRS	RSS
diploid	diploid

C	D
RRS	RSS
triploid	triploid

Figure 5.15

Questions 24 and 25 refer to the following information.

The formation of the fertile polyploid plant *Spartina townsendii* from *S. maritima* (2n = 56) and *S. alterniflora* (2n = 70) is thought to have involved the following stages.

1 vegetative growth of a sterile hybrid

2 complete non-disjunction (failure of chromosomes to separate during cell division)

3 fertilisation between gametes of two different species

24 If this is true, then these intermediate steps must have occurred in the order

   A 3, 1, 2   **B** 2, 1, 3   **C** 3, 2, 1   **D** 2, 3, 1

25 The chromosome complement of *Spartina townsendii* is

   A 2n = 63   **B** 2n = 91   **C** 2n = 98   **D** 2n = 126

Matching test part 1

Match the terms in list X with their descriptions in list Y.

list X
1 directional
2 disruptive
3 evolution
4 horizontal inheritance
5 natural selection
6 sexual selection
7 stabilising
8 vertical inheritance

list Y
a) form of selection that results in a population being split into two distinct groups, each with its own mean value for a trait
b) form of selection that leads to a reduction in genetic diversity without a change in a population's mean value for a trait
c) form of selection that results in a progressive shift in a population's mean value for a trait
d) non-random increase in frequency of DNA sequences which confer an advantage and increase the chance of survival
e) transfer of genetic sequences from one generation 'down' to the next by sexual or asexual reproduction
f) transfer of genetic material 'sideways' between contemporary members of a population
g) process of gradual change in a population of organisms over successive generations resulting from genomic variations
h) non-random increase in frequency of DNA sequences that increase the reproductive success of a species

Matching test part 2

Match the terms in list X with their descriptions in list Y.

list X
1 allopatric
2 founder effect
3 gene pool
4 genetic drift
5 isolating mechanism
6 neutral
7 selective advantage
8 speciation
9 species
10 sympatric

list Y
a) benefit gained by mutant organisms in an environment that suits them but not other members of the population
b) geographical, ecological or behavioural barrier that prevents genetic exchange between populations of a species
c) form of speciation where gene flow between populations of a species is prevented by a behavioural or ecological barrier
d) form of speciation where gene flow between populations of a species is prevented by a geographical barrier
e) generation of new biological species by evolution as a result of isolation, mutation and selection
f) form of genetic drift involving the isolation of a splinter group that possesses gene frequencies unrepresentative of the original population's gene pool
g) group of organisms that produce fertile offspring and share the same chromosome complement
h) random increase or decrease in the frequency of genetic sequences, particularly in small populations, that occurs due to sampling error
i) total of all the genomic sequences (alleles) present in the genotypes of the members of a species
j) type of mutation whose genetic sequence differs from the original but which still codes for the functional protein

➡

Multiple choice test

Choose the ONE correct answer to each of the following multiple choice questions.

1 The following diagram shows examples of genetic material being transferred in different directions by natural means (P = prokaryotes and E = eukaryotes). Which one is NOT correct?

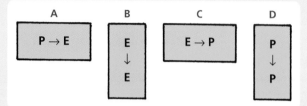

Figure 7.1

2 Experts believe that during the very early stages of evolution on Earth, rapid evolutionary change occurred in
A prokaryotes largely as a result of vertical gene transfer.
B eukaryotes largely as a result of vertical gene transfer.
C prokaryotes largely as a result of horizontal gene transfer.
D eukaryotes largely as a result of horizontal gene transfer.

3 Which of the following conditions that affect humans results from horizontal gene transfer?

	condition	cause
A	sickle cell trait	receipt of co-dominant alleles in genome
B	colour-blindness	presence in genome of recessive allele(s) of sex-linked gene
C	Huntington's chorea	inheritance of lethal dominant mutant allele
D	AIDS	infection by human immuno-deficiency virus (HIV)

Table 7.1

4 MRSA (methicillin-resistant *Staphylococcus aureus*) is a strain of bacterium that is resistant to several antibiotics as a result of receiving certain genomic sequences in
A plasmids from other bacteria by horizontal gene transfer.
B plasmids from a virus by vertical gene transfer.
C ribosomes from other bacteria by vertical gene transfer.
D ribosomes from a virus by horizontal gene transfer.

Questions 5 and 6 refer to the following graph which shows the effect of an antibiotic on two strains of a species of bacterium.

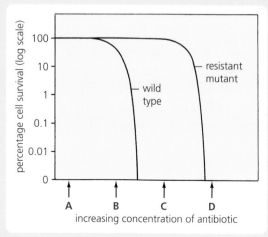

Figure 7.2

5 To which concentration of antibiotic are both strains of the bacterium sensitive?
6 Which concentration of antibiotic would be suitable for use in selecting the resistant mutant strain?

Questions 7, 8 and 9 refer to the following information and diagram of land snails.
The shell of the land snail shows variation in both colour and banding pattern. In order to construct a five-figure banding formula, bands are numbered from the top of the largest whorl as shown below. Zero (0) is used to represent the absence of a band and square brackets indicate the fusion of two bands.

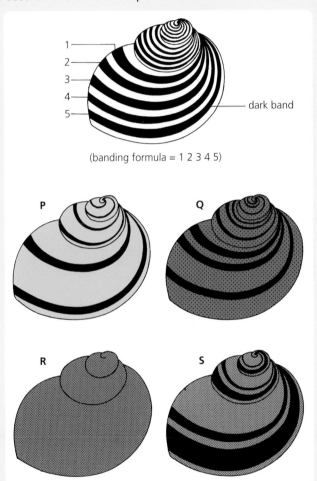

(banding formula = 1 2 3 4 5)

Figure 7.3

7 Which line in the table correctly identifies the banding formula of each of the three shells?

	shell		
	P	Q	S
A	02005	12040	003[45]
B	00305	12040	003[45]
C	02005	02305	030[45]
D	00305	02305	030[45]

Table 7.2

8 Thrushes (which have good colour vision) smash the shells of land snails against stones (anvils) in order to feed on the soft inner body. If snail types P, Q, R and S began in equal numbers in a habitat of grassland, which would be MOST likely to enjoy the greatest selective advantage?

A P   B Q   C R   D S

9 A survey of broken shells collected from thrush anvils among dead beech leaves in a woodland area was carried out. Predict which of the following sets of results was obtained.

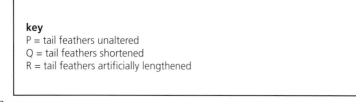

| | % number of broken shells of each type | | | |
	P	Q	R	S
A	13	33	1	5
B	11	1	34	6
C	5	1	14	32
D	6	21	20	5

Table 7.3

10 Over many generations within a population, a deleterious genetic sequence normally undergoes a
  A random reduction in frequency.
  C random increase in frequency.
  B non-random reduction in frequency.
  D non-random increase in frequency.

11 Which of the following gives TWO ways in which sexual selection may operate?
  A male to female competition for territory and female choice of mate
  B male to male competition for territory and female choice of mate
  C male to female competition for territory and male choice of mate
  D female to female competition for territory and male choice of mate

Questions 12 and 13 refer to the bar chart below and the information that follows.

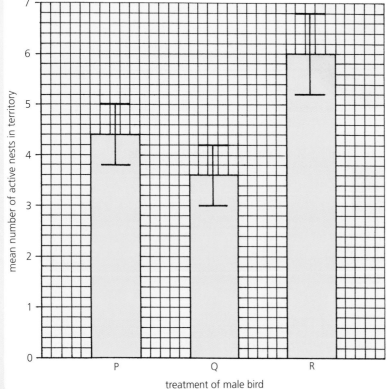

Figure 7.4

The male African widowbird develops long tail feathers during the breeding season and secures a territory where each female with whom he mates lays her eggs in a separate active nest. The bar chart represents the results of a sexual selection investigation into the effect of male tail length on the mean number of active nests established by females in a male territory.

12 With a 95% level of confidence, it can be stated that the minimum number of active nests present in the territory of a bird that has had its feathers artificially lengthened is

   A 3.0   **B** 3.8   **C** 5.2   **D** 6.0

13 Which of the following conclusions can be correctly drawn from the data?

   1 The number of active nests in the territory of a bird that received treatment P is significantly different from that of a bird that received treatment Q.

   2 The number of active nests in the territory of a bird that received treatment P is significantly different from that of a bird that received treatment R.

   3 The number of active nests in the territory of a bird that received treatment Q is significantly different from that of a bird that received treatment R.

   A 3 only   **B** 1 and 2   **C** 2 and 3   **D** 1, 2 and 3

Questions 14 and 15 refer to the following information and the graphs in the diagram below.

Several types of artificial selection can be employed to alter a crop plant's quantitative characteristics as follows.

1 Disruptive selection is practised when a crop is being developed for two different markets (e.g. barley with a low nitrogen content for brewing and barley with a high nitrogen content for livestock feed).

2 Stabilising selection is used for maintaining uniformity (e.g. crop height to suit harvesting machinery).

3 Directional selection is practised if increase in yield per plant is required.

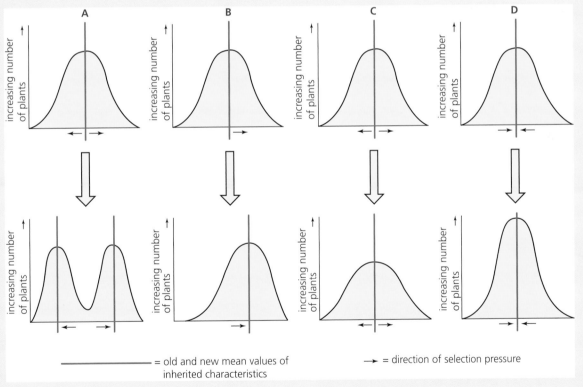

Figure 7.5

14 Which set of graphs represents stabilising selection?

15 Which set of graphs represents directional selection?

16 The sum total of all the different genetic sequences (alleles of genes) possessed by the members of an interbreeding population is known as the

   A gene frequency.   **B** gene code.   **C** gene flow.   **D** gene pool.

17 Which line in the table supplies the correct answers to blanks 1 and 2 in the following statement?

Genetic drift is the ___1___ change in frequency of genetic frequencies that occurs due to a non-representative sample of the ___2___ of the whole population being passed on to future generations.

	blank 1	blank 2
A	random	gametes
B	non-random	gametes
C	random	alleles
D	non-random	alleles

Table 7.4

18 The boxed diagram below represents the founder effect.

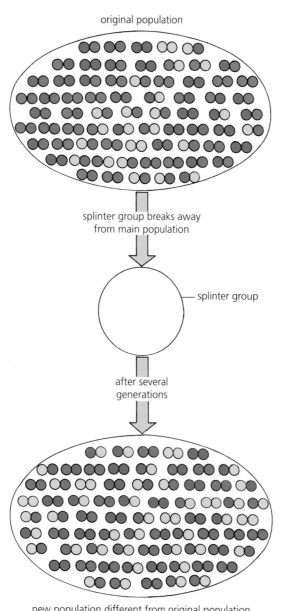

Figure 7.6

Which circle in the following diagram shows the composition of the splinter group?

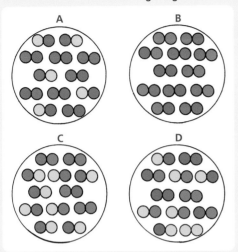

Figure 7.7

19 Which of the following statements is correct?
   A   Allopatric speciation is promoted by directional selection.
   B   Allopatric speciation is promoted by stabilising selection.
   C   Sympatric speciation is promoted by directional selection.
   D   Sympatric speciation is promoted by disruptive selection.

Questions 20 and 21 refer to the following table which gives details of some Scottish varieties of *Apodemus*, the long-tailed field mouse. These animals live on the islands shown on the accompanying map.

scientific name			location	average head and body length (mm)	average tail length (mm)	dorsal colour
genus	species	subspecies				
Apodemus	*hebridensis*	*hebridensis*	Lewis	95.5	87.8	wood brown
Apodemus	*hebridensis*	*tirae*	Tiree	102.5	84.2	reddish-brown
Apodemus	*hebridensis*	*hamiltoni*	Eigg	103.8	95.6	pale brown
Apodemus	*hebridensis*	*hirtensis*	St Kilda	110.9	105.5	peppery brown

Table 7.5

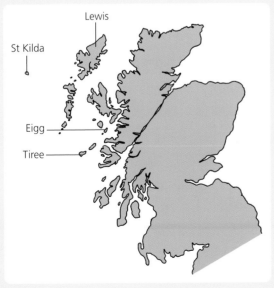

Figure 7.8

20 The isolating mechanism promoting speciation among the populations of *Apodemus* is an example of

    A a physiological barrier.
    B an ecological barrier.
    C a reproductive barrier.
    D a geographical barrier.

21 A long-tailed Hebridean field mouse was found to have a head and body length of 101.9 mm and a tail length of 85.1 mm. It is most likely that it was a native of

    A Lewis.
    B Tiree.
    C Eigg.
    D St Kilda.

22 The boxes in the following diagram give six stages that occur during the process of sympatric speciation.

> **1** occupation of a habitat by one large interbreeding population of one species
>
> **2** mutation
>
> **3** successful exploitation of an alternative ecological niche by a subpopulation
>
> **4** natural selection
>
> **5** isolation of the two populations caused by their behaviour
>
> **6** formation of a new species

Figure 7.9

Which of the following is the correct sequence?

    A 1, 3, 5, 2, 4, 6
    B 1, 5, 3, 2, 4, 6
    C 1, 3, 5, 4, 2, 6
    D 1, 5, 3, 4, 2, 6

Questions 23, 24 and 25 refer to the diagram below which shows the geographical distribution of five populations of a certain type of seabird.

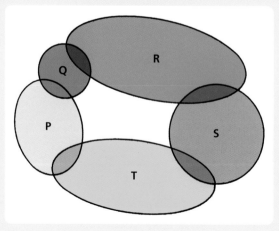

Figure 7.10

The following table shows the results of breeding experiments (where + = successful interbreeding and - = unsuccessful interbreeding).

cross	result
P x Q	+
P x R	-
P x S	-
P x T	-
Q x R	+
Q x S	-
Q x T	-
R x S	+
R x T	-
S x T	+

Table 7.6

23 How many different species are present?

  A 1   B 2   C 4   D 5

24 If population P became extinct, how many species would be present?

  A 1   B 2   C 3   D 4

25 If, on the other hand, population S became extinct, how many species would be present?

  A 1   B 2   C 3   D 4

8 Genomic sequencing

Matching test
Match the terms in list X with their descriptions in list Y.

list X
1 bioinformatics
2 conserved
3 divergence
4 domain
5 genome
6 genomics
7 model organism
8 molecular clock
9 neutral
10 pharmacogenetics
11 phylogenetic tree
12 phylogenetics
13 restriction endonuclease
14 sequencing
15 single nucleotide polymorphism

list Y
a) study of genomes related to personalised medicine
b) species important for research because it possesses genes equivalent to human genes responsible for inherited disorders
c) type of mutation whose genetic sequence differs from the original but which still codes for the functional protein
d) determining the order of bases on DNA fragments and the order of the fragments in a genome
e) type of enzyme that recognises a specific sequence of nucleotides on a DNA strand and cuts the DNA at such a site
f) a variation in DNA sequence that affects a single base pair in a DNA chain
g) application of statistics and computer science to analyse and compare genetic sequence data
h) study of evolutionary relatedness among different groups of organisms
i) separation of two groups in a phylogeny when their genomes acquire mutations and become different from one another
j) large group of living organisms such as bacteria, archaea or eukaryotes
k) use of a molecule that has accumulated mutations over time to construct an evolutionary timescale for groups possessing the molecule
l) term used to describe DNA sequences found to be very similar in the genomes of two organisms being compared
m) the entire genetic information of an organism encoded in the DNA of a complete set of its chromosomes
n) application of sequencing and computational procedures to analyse the genomes of organisms
o) branching diagram showing evolutionary relationships between groups of organisms based on their genomic differences

Multiple choice test
Choose the ONE correct answer to each of the following multiple choice questions.

1 The following diagram shows the DNA fragments which resulted from two copies of part of a gene, each cut by a different restriction endonuclease. The computer found five of the fragments to possess matching genetic sequences as indicated by the coloured regions.

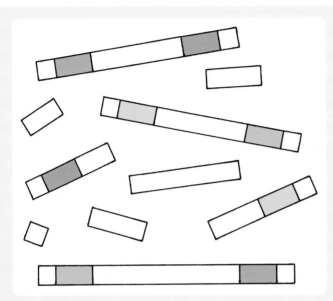

Figure 8.1

Which part of the diagram below indicates the correct genomic sequence?

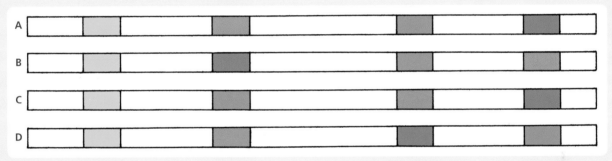

Figure 8.2

Questions 2 and 3 refer to the following information and to the possible answers that follow.

The DNA fragments shown in the accompanying diagram were formed during a type of sequencing where each fluorescent tag indicates the point on the strand where replication of complementary DNA was brought to a halt by a modified nucleotide.

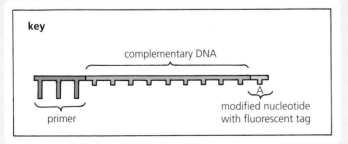

Figure 8.3

A GTAACTGCTA   B TCGTCAATGA   C CATTGACGAT   D AGCAGTTACT

2 Which is the sequence of bases in the complementary DNA strand?
3 Which is the sequence of bases in the original DNA strand?

Questions 4, 5 and 6 refer to the data about the four organisms in the table.

organism	length of genome (kb)	number of genes that code for protein	ratio of genome length to gene number	relative number of introns
1	1 650	1 500	[box W]	none
2	[box X]	4 200	1.2 : 1	none
3	6 999	[box Y]	1.5 : 1	a few
4	40 000	10 000	4 : 1	many

Table 8.1

(Note: 1 kilobase (kb) = 1 x 10³ bases)

4 Which value should have been entered in box W?
   A 1 : 1.1   B 0.91 : 1   C 1.1 : 1   D 150 : 1

5 Which line in the following table correctly shows the values that should have been inserted in boxes X and Y?

	X	Y
A	3 500	4 666
B	5 040	4 666
C	3 500	10 499
D	5 040	10 499

Table 8.2

6 Which organism is most likely to have the highest quantity of repetitive DNA in its genome?

A 1   B 2   C 3   D 4

7 The following table compares the genome of the puffer fish with two other common vertebrates.

organism	genome size (Mb)	approximate number of protein-coding genes
puffer fish	393	31 000
dog	2 400	19 300
mouse	3 400	23 000

Table 8.3

(Note: 1 megabase (Mb) = 1 x 10^6 bases)

The differences in the data are accounted for by the fact that compared with other vertebrates, the genome of the puffer fish has undergone a higher rate of

A deletions within chromosomes and possesses fewer introns.

B deletions within chromosomes and possesses fewer exons.

C duplication of genes and possesses more introns.

D duplication of genes and possesses more exons.

8 The degree of evolutionary relatedness among different groups of organisms based on comparisons of genomic sequences can be used to construct a phylogenetic tree. Which of the trees in the accompanying diagram correctly represents the data in the following table?

groups	genetic distance between groups (units)
CA and K	6
J and K	3
CA and L	6
M and N	5
CA and N	7

Table 8.4

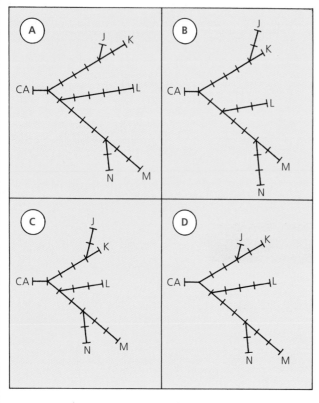

(Note: ⊢—⊣ = one unit of genetic change)

Figure 8.4

9 Which of the phylogenetic trees in the following diagram correctly represents the relationship between the three domains of living things based on genetic evidence?

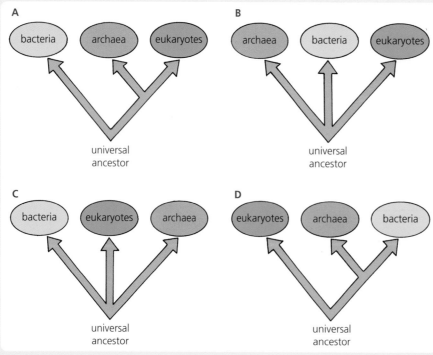

Figure 8.5

10 The source of the genetic evidence for the 'three domains' theory is derived from genetic material taken from many organisms which enables scientists to compare their

A nucleotide sequences of transfer RNA.

B anticodon sequences of transfer RNA.

C nucleotide sequences of ribosomal RNA.

D anticodon sequences of ribosomal RNA.

Questions 11 and 12 refer to the graph which shows the use of haemoglobin as a molecular clock.

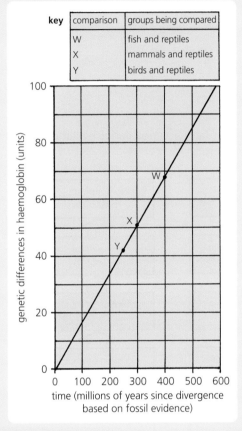

key	comparison	groups being compared
	W	fish and reptiles
	X	mammals and reptiles
	Y	birds and reptiles

Figure 8.6

11 Which line in the following table is correct?

	group being compared	point in time when groups diverged (millions of years ago)
A	birds and reptiles	200
B	mammals and reptiles	300
C	amphibians and reptiles	400
D	fish and reptiles	500

Table 8.5

12 The haemoglobin from two types of fish (carp and lamprey) was analysed.

The relative number of genetic differences by which their haemoglobin was found to differ was 86. This suggests that they shared a common ancestor and diverged around

A 200 million years ago.

B 300 million years ago.

C 400 million years ago.

D 500 million years ago.

13 The following table compares the three main domains of living things. Which line is correct?

	characteristic	domain		
		bacteria	archaea	eukaryotes
A	true nucleus bound by double membrane	+	-	+
B	membrane-enclosed organelles	-	-	+
C	introns	-	+	-
D	exons	+	-	+

Table 8.6

(Note: + = present, - = absent)

14 Which flow chart in the diagram shows the correct sequence of events that have occurred during evolution?

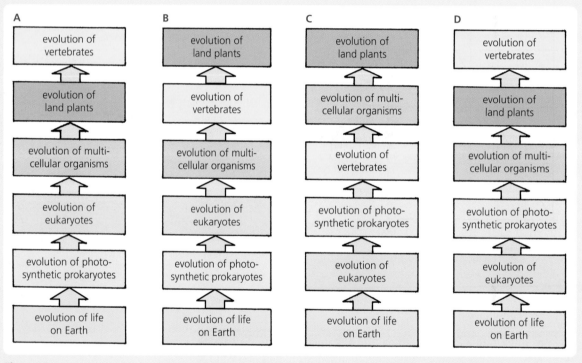

Figure 8.7

15 The table below gives the answers to the three blanks in the following sentence. Which line is correct?
In the future, knowledge of the genetic components of diseases and the use of personal ___1___ may lead to ___2___ and each medical treatment will be customised to suit the requirements of an individual's ___3___ .

	blank 1	blank 2	blank 3
A	genomics	metabolism	pharmacogenetics
B	metabolism	pharmacogenetics	genomics
C	genomics	pharmacogenetics	metabolism
D	metabolism	genomics	pharmacogenetics

Table 8.7

Metabolism and Survival

9 Metabolic pathways and their control

Matching test part 1

Match the terms in list X with their descriptions in list Y.

list X
1 active transport
2 alternative route
3 anabolism
4 carrier
5 catabolism
6 channel-forming
7 chloroplast
8 compartment
9 concentration gradient
10 diffusion
11 lysosome
12 metabolism
13 mitochondrion
14 reversible

list Y
a) difference that exists between two regions resulting in diffusion
b) step in a metabolic pathway that can operate in both a forward and backward direction
c) region of an organelle enclosed by membrane(s), enabling metabolic activity to be localised
d) movement of molecules or ions through a cell membrane from a region of lower concentration to a region of higher concentration
e) movement of molecules or ions from a region of higher concentration to a region of lower concentration
f) type of metabolic pathway that brings about the biosynthesis of complex molecules and requires energy
g) type of metabolic pathway that brings about the breakdown of complex molecules and releases energy
h) type of protein molecule in cell membrane containing a pore through which specific substances are able to diffuse
i) type of protein molecule that actively pumps specific ions into and/or out of a cell
j) membrane-bound organelle containing enzymes responsible for the citric acid cycle
k) membrane-bound organelle containing enzymes responsible for the Calvin cycle
l) membrane-bound organelle containing digestive enzymes
m) step(s) in a metabolic pathway that allow the regular steps to be bypassed
n) sum of all the biochemical reactions occurring within a living organism

Matching test part 2

Match the terms in list X with their descriptions in list Y.

list X
1 activation energy
2 active site
3 affinity
4 catalyst
5 end product
6 enzyme
7 induced fit
8 orientation
9 reaction rate
10 specificity
11 substrate
12 transition

list Y
a) substance that increases the rate of a chemical reaction and remains unaltered by the reaction
b) complementary relationship of a molecular structure allowing an enzyme to combine with one type of substrate only
c) substance formed as a result of an enzyme acting on its substrate
d) substance upon which an enzyme acts, resulting in the formation of an end product
e) energy needed to break the chemical bonds in the reactants in a chemical reaction
f) region of an enzyme molecule where the complementary surface of its substrate molecule becomes attached
g) degree of chemical attraction between reactant molecules
h) state of reactant molecules that have absorbed enough energy to break their bonds and allow the reaction to occur
i) protein made by living cells that acts as a biological catalyst
j) state of close molecular contact resulting from change in shape of an enzyme's active site to accommodate its substrate
k) way in which molecules of two reactants are held together as determined by the shape of the enzyme's active site
l) amount of chemical change that occurs per unit of time

Multiple choice test part 1

Choose the ONE correct answer to each of the following multiple choice questions.

1   Which of the following terms could NOT be used to describe correctly the type of pathway represented by the blue arrows in the following diagram?

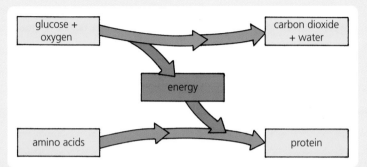

Figure 9.1

A   metabolic   B   anabolic   C   biochemical   D   catabolic

2   Which line in the table below would correctly refer to the compartment in the mitochondrion shown in the following diagram if the folds in its inner membrane were absent?

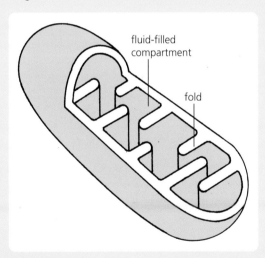

Figure 9.2

	relative surface area to volume ratio	relative level of metabolic activity
A	lower	decreased
B	higher	decreased
C	lower	increased
D	higher	increased

Table 9.1

3 The following diagram allows a comparison to be made of the surface area to volume ratio of two cubes. Which line in the table below is correct?

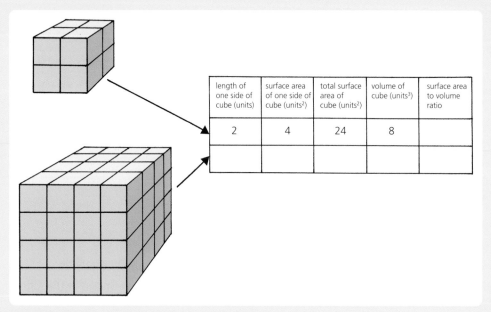

length of one side of cube (units)	surface area of one side of cube (units²)	total surface area of cube (units²)	volume of cube (units³)	surface area to volume ratio
2	4	24	8	

Figure 9.3

	surface area to volume ratio	
	smaller cube	larger cube
A	3:1	2:3
B	3:1	3:2
C	1:3	2:3
D	1:3	3:2

Table 9.2

4 The following diagram shows the fluid-mosaic model of the structure of a cell membrane.

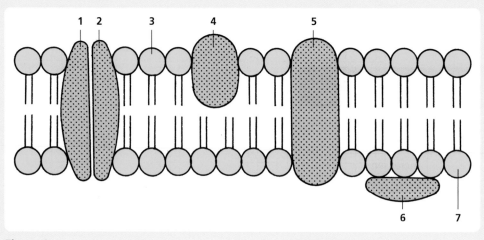

Figure 9.4

Which pair of numbered structures is correctly identified in the accompanying table?

	protein	phospholipid
A	1	7
B	2	4
C	3	7
D	5	6

Table 9.3

5 It would NOT be correct to say that protein present in the cell membrane can act as
  A an enzyme that catalyses a step in a biochemical reaction in the membrane.
  B a channel-forming molecule that allows molecules to diffuse through the membrane.
  C a continuous waterproof layer that protects the cell and its organelles.
  D a carrier molecule that actively pumps molecules into and out of the cell.

6 The following diagram shows ways in which molecules may move into and out of a respiring animal cell. Which arrow could be the diffusion of carbon dioxide molecules?

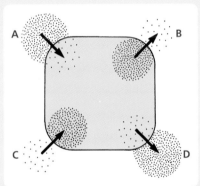

Figure 9.5

7 Active transport is the movement of ions or molecules across a cell membrane from a
  A low to a high concentration along a concentration gradient.
  B high to a low concentration along a concentration gradient.
  C low to a high concentration against a concentration gradient.
  D high to a low concentration against a concentration gradient.

8 The following diagram represents active transport of sodium and potassium ions through a cell membrane by the action of carrier molecules acting as pumps.

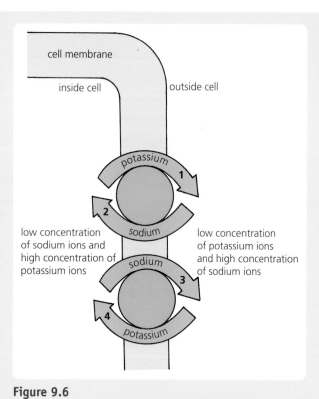

Figure 9.6

Which line in the table below correctly refers to the two arrows at which active transport is occurring?

	active transport of potassium ions	active transport of sodium ions
A	1	2
B	4	2
C	1	3
D	4	3

Table 9.4

Questions 9 and 10 refer to the following graphs.

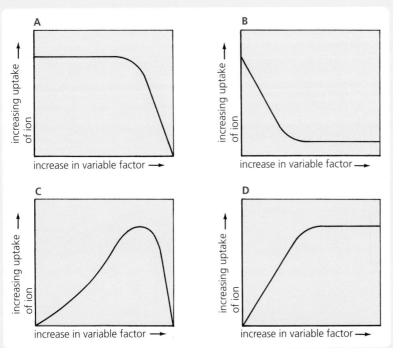

Figure 9.7

9 Which graph represents the rate of active uptake of a type of ion by a living cell in response to increasing oxygen concentration?

10 Which graph represents the rate of active uptake of a type of ion by a living cell in response to increasing temperature?

11 The diagram below shows the results of an analysis of cell sap from a marine plant and the surrounding sea water.

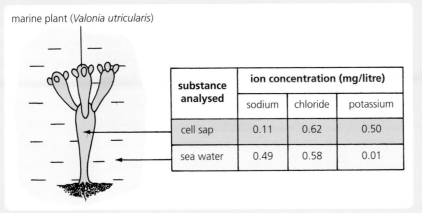

marine plant (*Valonia utricularis*)

substance analysed	ion concentration (mg/litre)		
	sodium	chloride	potassium
cell sap	0.11	0.62	0.50
sea water	0.49	0.58	0.01

Figure 9.8

From these data it can be concluded that this plant
A accumulates all three types of ion from sea water.
B holds chloride ions at a concentration lower than sea water.
C selects and internally accumulates sodium ions only.
D can differentiate between sodium and potassium ions.

12 The following graph shows the changes in ionic concentrations of culture solutions in which barley roots were grown for two days.

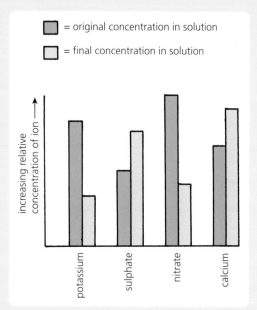

■ = original concentration in solution

□ = final concentration in solution

Figure 9.9

Which of the following ions were BOTH taken up by the plant?

A calcium and sulphate   B sulphate and potassium

C potassium and nitrate   D nitrate and calcium

13 Which of the following types of subcellular structure is ALWAYS associated with the active transport of ions across a cell membrane?

A ribosome   B mitochondrion

C lysosome   D nucleus

14 The table below shows the outcome of an investigation into the uptake of bromide ions by a plant.

time from start of experiment in minutes	units of bromide ions taken up by plant tissue under the following conditions		
	sugar absent, oxygen present	sugar present, oxygen absent	sugar and oxygen present
0	0	0	0
30	0	30	100
60	0	50	150
90	0	70	180
120	0	70	200

Table 9.5

These results indicate that uptake of bromide ions

A is an active process requiring energy.

B occurs during aerobic respiration only.

C requires a temperature suitable for enzymes to act.

D stops in the absence of oxygen.

15 Which of the following graphs represents a chemical reaction controlled by a catalyst?

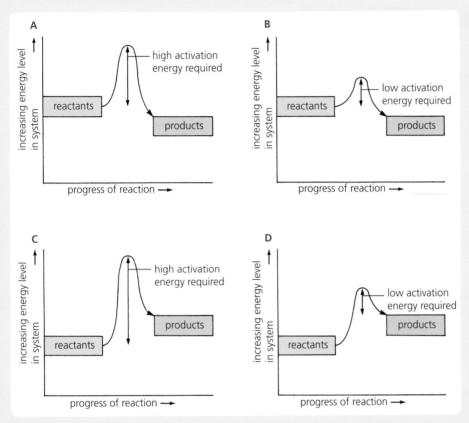

Figure 9.10

16 The experiment shown in the following diagram was set up to investigate the effect of catalase on the breakdown of hydrogen peroxide.

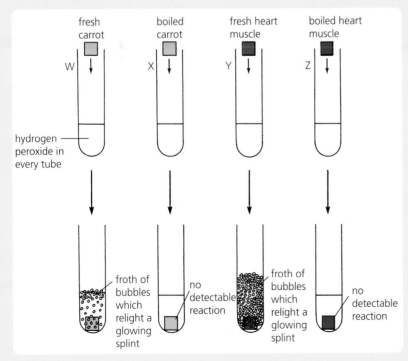

Figure 9.11

Which line in the table below is correct?

	substrate	end product
A	carrot/heart muscle	carbon dioxide
B	hydrogen peroxide	carbon dioxide
C	carrot/heart muscle	oxygen
D	hydrogen peroxide	oxygen

Table 9.6

Questions 17 and 18 refer to the following diagram which shows four stages that occur during an enzyme-controlled reaction.

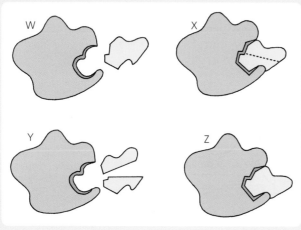

Figure 9.12

17 Which of the following indicates the correct sequence in which the four stages would occur if the enzyme promotes the building-up of a complex molecule from simpler ones?

A Y → X → Z → W   B W → Z → X → Y   C Y → Z → X → W   D W → X → Z → Y

18 Which stage(s) illustrate a state of induced fit between the enzyme and its substrate?

A X only   B Z only   C X and Z   D W, X and Z

19 The following graph charts the effect of increasing substrate concentration on the rate of an enzyme-controlled reaction where the concentration of enzyme is limited.

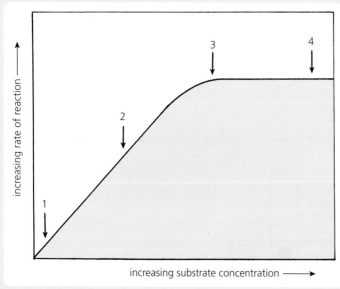

Figure 9.13

The concentration of substrate is the limiting factor at point(s)

A 1 only. **B** 1 and 2. **C** 3 only. **D** 3 and 4.

20 The following diagram shows a metabolic pathway where each encircled letter represents a metabolite.

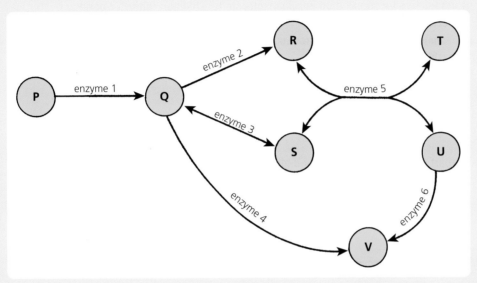

Figure 9.14

By which alternative route could a supply of metabolite V be obtained if enzyme 4 became inactive?

A Q → enzyme 2 → R → enzyme 5 → U → enzyme 6 → V   **B** Q → enzyme 3 → S → enzyme 6 → U → enzyme 5 → V
C S → enzyme 3 → Q → enzyme 2 → R → enzyme 6 → U   **D** S → enzyme 5 → T → enzyme 6 → R → enzyme 5 → U

Matching test part 3
Match the terms in list X with their descriptions in list Y.

list X
1 enzyme induction
2 inducer
3 operon
4 regulator gene
5 repressor
6 structural gene
7 transformation

list Y
a) alteration of a cell's genotype using a genetically modified plasmid
b) combination of an operator gene and one or more structural genes
c) region of DNA which codes for a repressor molecule
d) molecule which prevents a repressor molecule from combining with an operator gene
e) region of DNA which codes for a functional protein such as an enzyme
f) process by which an operator gene becomes free and an operon is able to code for an enzyme
g) molecule coded for by a regulator gene which can combine with an operator gene

Matching test part 4
Match the terms in list X with their descriptions in list Y.

list X
1 activator
2 active site
3 allosteric site
4 competitive
5 end-product inhibition

list Y
a) process by which a metabolite at a later stage in a pathway builds up and prevents the activity of an enzyme controlling an earlier stage
b) substance that acts on an enzyme as an activator or an inhibitor
c) regulatory molecule that becomes attached to an enzyme molecule holding it in its active form
d) type of inhibitor that becomes attached to a non-active site on an enzyme and changes the enzyme's molecular shape
e) region on an enzyme molecule to which the complementary surface of the substrate molecule becomes attached

list X
6 inhibitor
7 non-competitive
8 regulatory molecule
9 signal molecule

list Y
f) type of inhibitor with a molecular structure similar to an enzyme's substrate enabling it to become attached to the enzyme's active site
g) non-active location on an enzyme molecule to which an inhibitor or an activator may become attached
h) chemical substance (often from outside a cell) which exerts control over a metabolic pathway within the cell
i) regulatory molecule that halts or decreases the rate of an enzyme-controlled reaction

Multiple choice test part 2

Choose the ONE correct answer to each of the following multiple choice questions

1 The following diagram shows part of a metabolic pathway under genetic control.

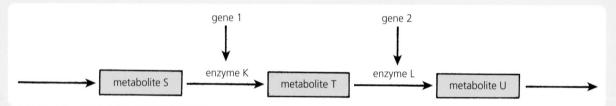

Figure 9.15

Which line in the table below correctly summarises the effect that a mutation could have on this pathway?

	site of major genetic fault	substance which accumulates as a result of an error of metabolism
A	1	S
B	1	T
C	2	L
D	2	U

Table 9.7

Questions 2, 3 and 4 refer to the following diagram which shows a possible arrangement of the genes involved in the induction of the enzyme β-galactosidase in the bacterium *Escherichia coli*.

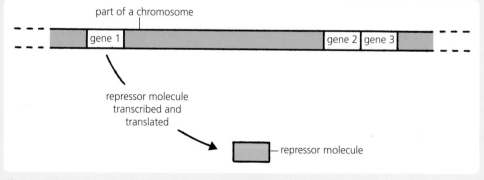

Figure 9.16

2 In the absence of lactose, the repressor molecule combines with gene 2 and, as a result, gene 3 remains 'switched off'. Which line in the following table indicates the correct identity of the three genes?

	operator gene	structural gene	regulator gene
A	2	3	1
B	1	3	2
C	3	2	1
D	2	1	3

Table 9.8

3 In this system, the operon consists of
   A gene 2 only.   B gene 3 only.   C genes 2 and 3.   D genes 1, 2 and 3.

4 Which of the following situations would arise if lactose became available to the cell?

	gene 1	gene 2	gene 3
A	-	-	+
B	-	+	-
C	+	-	+
D	+	+	+

Table 9.9
(Note: + = 'switched on', - = 'switched off')

Questions 5 and 6 refer to the following information, diagram and table of possible answers.

Tryptophan is an amino acid needed for the synthesis of proteins. Situation 1 in the diagram shows a set of circumstances where a structural gene remains switched on in a cell of *Escherichia coli*. Under different circumstances, the series of events shown in situation 2 is thought to occur. This brings about the repression of the synthesis of an enzyme.

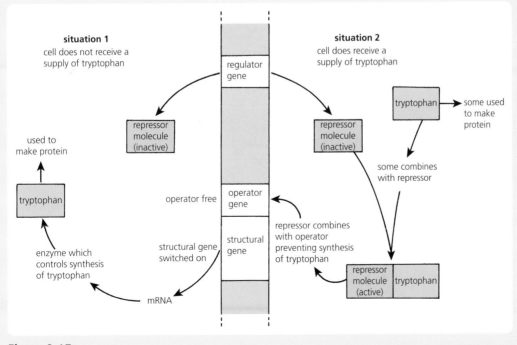

Figure 9.17

	regulator gene	operator gene	structural gene
A	on	blocked	off
B	off	blocked	off
C	off	free	on
D	on	free	on

Table 9.10

5 Which answer refers to the state of the genes in a cell of *Escherichia coli* grown in nutrient broth lacking tryptophan?

6 Which answer refers to the state of the genes in a cell of *Escherichia coli* cultured in nutrient broth containing tryptophan?

7 The diagram shows an experiment set up to investigate the lac operon of *Escherichia coli*. ONPG is a colourless synthetic chemical which can be broken down by the enzyme β-galactosidase to form yellow compounds.

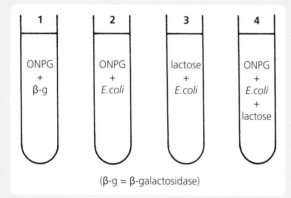

(β-g = β-galactosidase)

Figure 9.18

After an hour at 35°C a strong yellow colour will be formed in

A tube 4 only.   B tubes 1 and 4.   C tubes 1, 2 and 4   D tubes 2, 3 and 4.

Questions 8, 9 and 10 refer to the following diagram which represents the last four stages in a metabolic pathway in the fungus *Neurospora*.

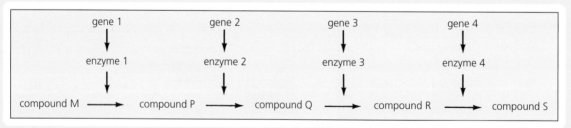

Figure 9.19

8 A mutant strain of the fungus is found to accumulate compound Q as a result of its metabolism. The gene which has undergone a mutation in this strain is

A 1   B 2   C 3   D 4

9 Wild-type *Neurospora* can grow on minimal medium (sucrose, mineral salts and one vitamin) but mutant strains suffering a metabolic block are unable to do so. In an experiment the mutant strain referred to in question 8 was subcultured onto the following plates.

Plate

P = minimal medium + substance P   **Q** = minimal medium + substance Q

R = minimal medium + substance R   **S** = minimal medium + substance S

This mutant strain would grow successfully on BOTH plates

A P and Q.   B Q and R.   C R and S.   D S and P.

10 A different mutant strain was found to grow successfully on plate S (minimal medium + substance S) but on no other. The enzyme that this strain of *Neurospora* fails to make is

A 1 B 2 C 3 D 4

11 The following diagram shows the action of an enzyme on its substrate.

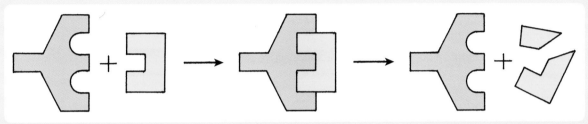

Figure 9.20

Which of the diagrams below shows how a competitive inhibitor brings about its effect on this system?

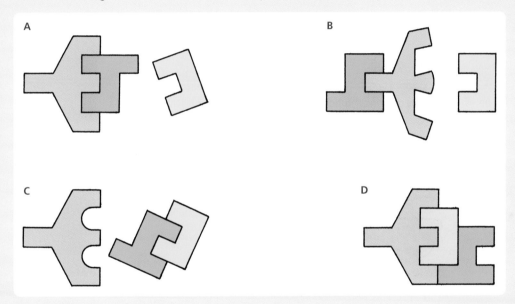

Figure 9.21

Questions 12 and 13 refer to the following graph which shows the effect of increasing substrate concentration on the rate of an enzyme-catalysed reaction affected by a limited amount of competitive inhibitor (and using a limited amount of enzyme).

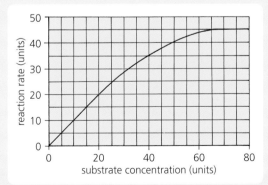

Figure 9.22

12 An increase in substrate concentration from 20 to 40 units brought about a percentage increase in reaction rate of

A 15   B 35   C 55   D 75

13 At 70 units of substrate the active sites on the enzyme molecules would be occupied by

A mostly inhibitor molecules.

B mostly substrate molecules.

C equal numbers of inhibitor and substrate molecules.

D neither inhibitor nor substrate molecules.

14 The following diagram shows an enzyme in its active and inactive forms.

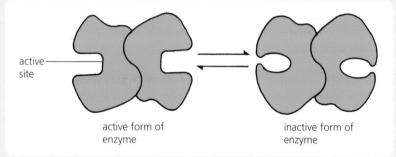

active site

active form of enzyme   inactive form of enzyme

Figure 9.23

Which of the diagrams below shows the action of a non-competitive inhibitor on this enzyme?

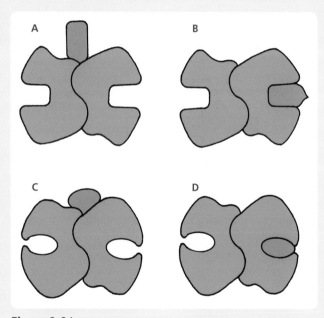

A

B

C

D

Figure 9.24

15 The red arrows in the following diagram of a metabolic pathway represent inhibition by an end product.

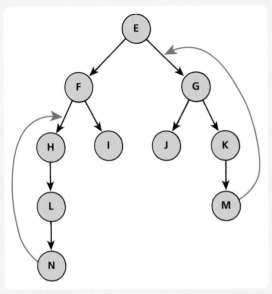

Figure 9.25

In the presence of high concentrations of metabolites M and N, which of the following chemical reactions would continue to proceed as before?

A E → G   B F → I   C G → J   D F → H

10 Cellular respiration

Matching test part 1

Match the terms in list X with their descriptions in list Y.

list X

1 ADP
2 anabolism
3 ATP
4 ATP synthase
5 catabolism
6 cellular respiration
7 dehydrogenase
8 energy transfer
9 glucose
10 phosphorylation
11 P_i

list Y

a) enzyme-controlled process by which a phosphate group is added to a molecule often making it more reactive

b) role played by ATP between energy-releasing and energy-consuming reactions

c) enzyme that removes hydrogen ions and high energy electrons from the respiratory substrate

d) respiratory substrate that provides energy for the regeneration of ATP

e) energy-consuming metabolic process often in the form of a biosynthetic pathway

f) general name for a metabolic process involving the breakdown of complex to simpler molecules normally releasing energy

g) low energy molecule composed of adenosine and two phosphate groups

h) inorganic phosphate group needed to make ATP from ADP

i) high energy molecule composed of adenosine and three phosphate groups

j) series of metabolic pathways which release energy from food allowing ATP to be regenerated

k) enzyme that catalyses the synthesis of ATP

Matching test part 2

Match the terms in list X with their descriptions in list Y.

list X

1 acetyl coenzyme A
2 carbon dioxide
3 citrate
4 citric acid cycle
5 electron transport chain
6 energy investment
7 energy payoff
8 ethanol
9 fermentation
10 glycogen and starch
11 glycolysis
12 inner mitochondrial membrane
13 lactate
14 NAD and FAD
15 oxaloacetate
16 oxygen
17 pyruvate
18 water

list Y

a) the final hydrogen acceptor which combines with hydrogen ions (and electrons) to form water

b) product of fermentation in animal cells

c) compound formed when glucose undergoes glycolysis

d) product of fermentation in plant cells

e) metabolite formed from oxaloacetate and an acetyl group

f) coenzyme molecules that act as hydrogen acceptors

g) location of electron transport chains in cells

h) metabolic pathway which breaks down glucose into pyruvate

i) first phase of glycolysis which requires ATP

j) gaseous product of aerobic respiration (and fermentation in plant cells)

k) compound formed from pyruvate and coenzyme A in the presence of oxygen

l) circular metabolic pathway of stages controlled by enzymes that remove hydrogen ions from the respiratory substrate

m) form of cellular respiration involving partial breakdown of glucose in the absence of oxygen

n) complex carbohydrates which can be broken down to glucose for use as the respiratory substrate

o) second phase of glycolysis which generates ATP

p) final product of aerobic respiration when oxygen combines with hydrogen

q) metabolite in the citric acid cycle which combines with an acetyl group to form citrate

r) group of protein molecules in a mitochondrial membrane which make energy available to pump hydrogen ions across the membrane

Multiple choice test

Choose the ONE correct answer to each of the following multiple choice questions.

1   Which of the following diagrams BEST represents the structure of a molecule of ATP (adenosine triphosphate)?

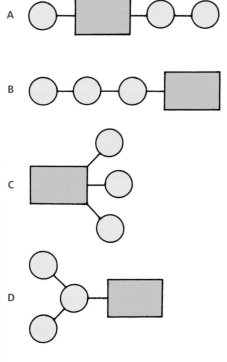

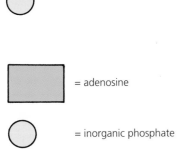

■ = adenosine

● = inorganic phosphate

Figure 10.1

2   Which of the following equations represents the regeneration of ATP from its components?

A $ADP + P_i \xrightarrow{\text{energy taken in}} ATP$

B $ADP + P_i + P_i \xrightarrow{\text{energy taken in}} ATP$

C $ADP + P_i \xrightarrow{\text{energy released}} ATP$

D $ADP + P_i + P_i \xrightarrow{\text{energy released}} ATP$

Questions 3 and 4 refer to the following diagram of tissue respiration.

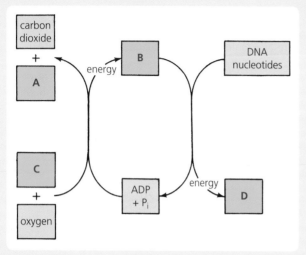

Figure 10.2

3 Which box represents the correct position of ATP in this scheme?

4 Which box represents the correct position of glucose in this scheme?

5 Which of the following is an example of anabolism?

  A deamination of amino acids to form urea   **B** digestion of starch by enzymes

  C formation of protein from amino acids   **D** oxidation of glucose during respiration

6 One mole of glucose releases 2880 kJ of energy. During aerobic respiration 44% of this is used to generate ATP. Therefore the number of kilojoules per mole of glucose transferred to ATP is

  A 126.7   **B** 161.3   **C** 1267.2   **D** 1612.8

7 The phosphorylation of glucose during glycolysis results in the production of

  A low energy glucose-6-phosphate and ATP.   **B** low energy glucose-6-phosphate and ADP.

  C high energy glucose-6-phosphate and ATP.   **D** high energy glucose-6-phosphate and ADP.

Questions 8 and 9 refer to the experiment shown in the diagram below. It was set up to investigate if glucose-1-phosphate (G-1-P), a phosphorylated form of glucose, is more reactive than glucose, the non-phosphorylated form. Starch-free potato extract contains the enzyme phosphorylase which promotes the synthesis of starch. The results were obtained by adding iodine solution to the cavities at 3-minute intervals.

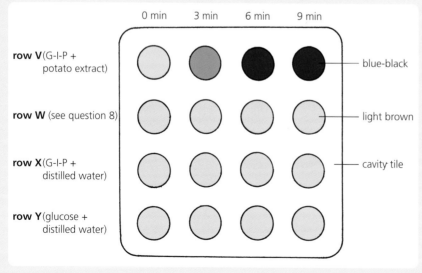

Figure 10.3

8 At the start of the experiment the cavities in row W should have received
  A glucose + potato extract.
  B glucose-1-phosphate + glucose.
  C glucose + iodine solution.
  D potato extract + distilled water.

9 If row X had not been set up as a negative control, then it would be valid to suggest that
  A G-1-P would have become phosphorylated in the presence of distilled water alone.
  B starch would have been formed whether or not the glucose was phosphorylated.
  C G-1-P would have become starch whether or not phosphorylase was present.
  D glucose would have become phosphorylated in the presence of phosphorylase.

10 Which line in the table below correctly refers to the effect that glycolysis of one molecule of glucose has on the number of molecules of ATP involved?

| | phase of glycolysis | | | |
| | energy investment | | energy payoff | |
	2 ATP used up	2 ATP generated	4 ATP used up	4 ATP generated
A		✓		✓
B		✓	✓	
C	✓		✓	
D	✓			✓

Table 10.1

Questions 11 and 12 refer to the following diagram which shows the metabolic pathways leading to and including the citric acid cycle.

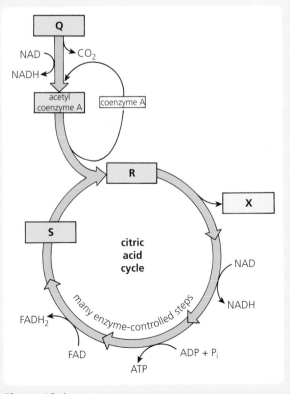

Figure 10.4

11 Which line in the following table indicates the correct identity of metabolites Q, R and S?

	Q	R	S
A	citrate	pyruvate	oxaloacetate
B	pyruvate	citrate	oxaloacetate
C	oxaloacetate	citrate	pyruvate
D	pyruvate	oxaloacetate	citrate

Table 10.2

12 The substance in box X should be

  A NADH.   B ATP.   C CO_2.   D water.

13 The enzymes required for the citric acid cycle in a cell are located in the

  A cytoplasmic fluid surrounding each mitochondrion.
  B outer membrane of each mitochondrion.
  C intermembrane space in each mitochondrion.
  D central matrix of each mitochondrion.

Questions 14, 15 and 16 refer to the following diagram. It represents a section through the inner membrane of a mitochondrion and shows the processes involved in energy transfer.

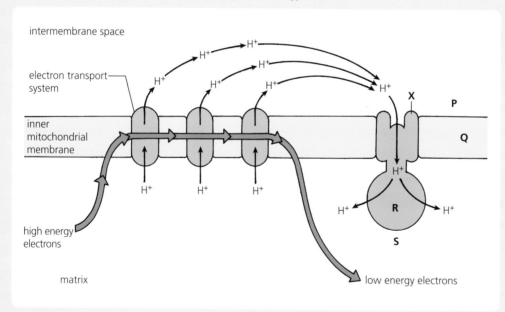

Figure 10.5

14 The region containing the highest concentration of hydrogen ions would be

  A P.   B Q.   C R.   D S.

15 The function of molecule X is to

  A promote energy release from ATP.   B act as an electron transport chain.
  C pump hydrogen ions out of the matrix.   D catalyse the synthesis of ATP.

16 The final electron acceptor in this system is

  A water.   B oxygen.   C ADP.   D NAD.

17 The enzymes that remove hydrogen ions from the respiratory substrate at several steps in glycolysis and the citric acid cycle are called

  A decarboxylases.   B phosphorylases.   C dehydrogenases.   D synthases.

18 In an investigation into aerobic respiration in yeast cells, four test tubes were set up as indicated in the table below (where ✓ = present and X = absent).

Resazurin dye is a chemical which changes colour upon gaining hydrogen as follows:

blue ⟶ pink ⟶ colourless
(lacks hydrogen)   (some hydrogen gained)   (much hydrogen gained)

substance	test tube			
	1	**2**	**3**	**4**
live yeast	✓	X	✓	X
dead yeast	X	X	X	✓
glucose solution	X	✓	✓	✓
resazurin dye	✓	✓	✓	✓

Table 10.3

The disappearance of resazurin's blue colour will
A occur in tube 3 more quickly than in tube 1.
B occur in tube 2 more quickly than in tube 1.
C fail to occur in both tubes 1 and 2.
D fail to occur in both tubes 3 and 4.

19 During aerobic respiration of one molecule of glucose, most ATP is synthesised during
A the citric acid cycle.   **B** electron transport.
C glycolysis.   **D** the breakdown of oxaloacetate.

Questions 20 and 21 refer to the following diagram. It shows the relationship between carbohydrate and two other classes of food which can act as respiratory substrates.

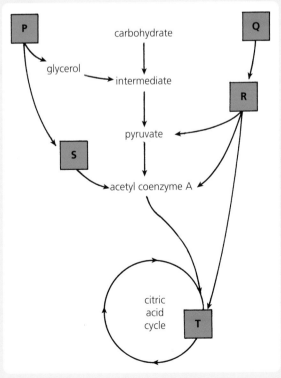

Figure 10.6

20 Which of the following boxes represents fats?

   A P   **B** Q   **C** S   **D** T

21 Which of the following boxes represents amino acids?

   A P   **B** R   **C** S   **D** T

Questions 22 and 23 refer to the following diagram. It shows the apparatus set up to investigate the use of sucrose as a respiratory substrate by yeast.

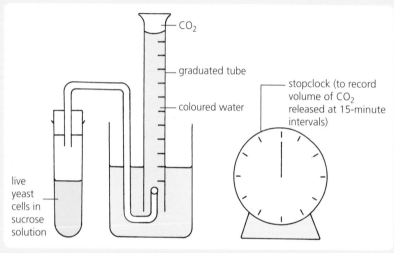

Figure 10.7

22 A suitable control would be an exact copy of the apparatus except that the test tube would contain

   A live yeast cells and glucose solution.   **B** live yeast cells and sucrose solution.

   C dead yeast cells and glucose solution.   **D** dead yeast cells and sucrose solution.

23 Which line in the table below is correct?

	dependent variable	independent variable
A	volume of CO_2	time
B	volume of CO_2	concentration of sucrose solution
C	concentration of sucrose solution	time
D	time	volume of CO_2

Table 10.4

Questions 24 and 25 refer to the following possible answers.

   A ethanol + CO_2 + ATP   **B** ethanol + ADP

   C lactic acid + CO_2 + ADP   **D** lactic acid + ATP

24 Which answer identifies the end products resulting from fermentation in a root cell from water-logged soil?

25 Which answer identifies the end products resulting from fermentation in mammalian tissue?

11 Metabolic rate

Matching test
Match the terms in list X with their descriptions in list Y.

list X
1. atrium
2. capillary bed
3. complete
4. cyanobacteria
5. double
6. incomplete
7. maximum oxygen uptake
8. metabolic rate
9. parabronchus
10. probe
11. single
12. ventricle

list Y
a) prokaryotes thought to be the first organisms to be able to photosynthesise
b) one of many tiny channels in the lungs of a bird which promote efficient gas exchange
c) circulatory system in which blood passes through the heart once for each complete circuit of the body
d) circulatory system in which blood passes through the heart twice for each complete circuit of the body
e) double circulatory system whose one ventricle may allow some mixing of oxygenated and deoxygenated blood
f) double circulatory system with two ventricles and no mixing of oxygenated and deoxygenated blood
g) region of the body containing a network of tiny blood vessels which offer resistance to the flow of blood
h) heart chamber which receives blood from capillary beds
i) heart chamber which pumps blood out of the heart to capillary beds
j) physiological measurement which acts as an indicator of a person's cardiovascular fitness
k) sensor used to measure changes in factors such as temperature in a respirometer
l) measure of oxygen consumption, carbon dioxide production or heat production per unit time by an organism

Multiple choice test
Choose the ONE correct answer to each of the following multiple choice questions.

Questions 1, 2 and 3 refer to the following experiment which was set up to measure a grasshopper's rate of metabolism. After 30 minutes the coloured liquid in the experiment was returned to its original level by depressing the syringe plunger from point X to point Y.

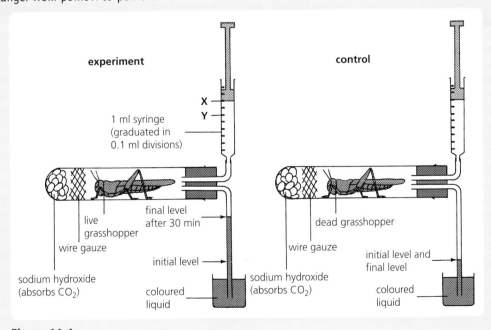

Figure 11.1

1 The rise in level of coloured liquid indicates that
   A the grasshopper is giving out carbon dioxide.
   B heat energy is released by the grasshopper.
   C the grasshopper is taking in oxygen.
   D the sodium hydroxide is absorbing oxygen.

2 From this experiment it can be concluded that the grasshopper's rate of
   A oxygen consumption is 2.0 ml/hour.
   B oxygen consumption is 0.4 ml/hour.
   C carbon dioxide output is 0.2 ml/hour.
   D carbon dioxide output is 4.0 ml/hour.

3 Which of the following procedures would improve the reliability of the result?
   A replacing the dead grasshopper in the control tube with glass beads
   B using a variety of insects with a period of acclimatisation allowed between readings
   C repeating the experiment with the same grasshopper and calculating an average
   D pooling class results where each group uses an adult locust

4 If an animal of mass 7 g consumes 3.5 cm^3 of oxygen in 25 minutes, then its metabolic rate in cm^3 oxygen used per gram of body tissue per minute is
   A 0.02   B 0.08   C 0.20   D 0.80

5 In which of the following pairs do BOTH animals possess an incomplete double circulatory system?
   A rabbit and newt   B toad and lizard   C pigeon and monkey   D snake and eagle

6 In a single circulatory system the route taken by blood on leaving the heart is
   A gills → body → heart.   B body → gills → heart.   C gills → heart → body.   D body → heart → gills.

Questions 7 and 8 refer to the following graph of the vascular pressures at various points in a dogfish's circulation.

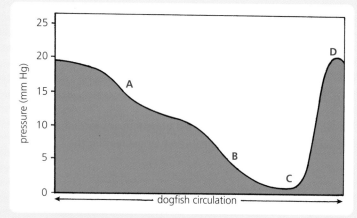

Figure 11.2

7 At which point on the graph is blood leaving the heart?
8 At which point on the graph is blood leaving the gills?

9 The following diagram shows a circulatory system. Which vessel contains oxygenated blood at high pressure?

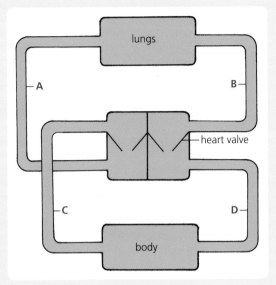

Figure 11.3

Questions 10 and 11 refer to the following diagram of part of a frog's circulatory system.

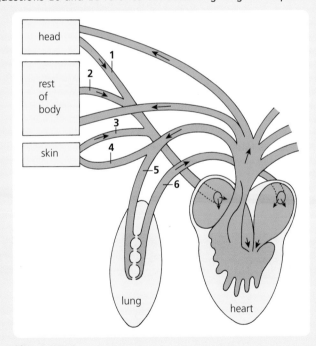

Figure 11.4

10 When a frog is active on land, blood with the highest oxygen content would be found in vessel

  A 2   **B** 4   **C** 5   **D** 6

11 When a frog is completely submerged in water, most oxygen enters the system by vessel

  A 1   **B** 2   **C** 3   **D** 6

12 The accompanying diagram shows a bird's breathing system.

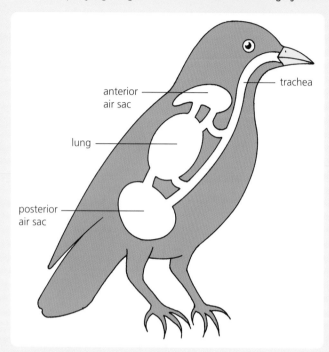

Figure 11.5

Which of the following statements is NOT correct?
A During inhalation the posterior air sacs fill up with fresh air.
B During inhalation the anterior air sacs receive fresh air from the lungs.
C During exhalation the posterior air sacs empty, forcing fresh air into the parabronchi.
D During exhalation the anterior air sacs empty, forcing stale air out of the system.

13 A person with a red blood cell count of 5×10^{12} l^{-1} at sea level was found to have a count of 6×10^{12} l^{-1} after living for two months at an altitude of 4 km above sea level. What percentage increase in red blood cell numbers had occurred?

A 1   B 17   C 20   D 30

14 Which line in the following table correctly applies to a seal during a deep-water dive?

	heart rate		state of lungs		buoyancy	
	increased	decreased	partially collapsed	greatly inflated	increased	decreased
A	X	✓	✓	X	X	✓
B	X	✓	X	✓	✓	X
C	✓	X	✓	X	X	✓
D	✓	X	X	✓	✓	X

Table 11.1
(Note: ✓ = yes, X = no)

15 The boxed statements in the following diagram give the stages involved in measuring maximum oxygen uptake. What is their correct sequence?

> ① person begins exercise on an ergometer

> ② maximum oxygen uptake determined when oxygen consumption stays steady despite an increase in workload

> ③ person is fitted with oxygen and carbon dioxide analyser

> ④ workload is gradually increased incrementally from moderate to maximum

Figure 11.6

A 1,3,2,4   B 3,1,2,4   C 1,3,4,2   D 3,1,4,2

12 Metabolism in conformers and regulators

Matching test
Match the terms in list X with their descriptions in list Y.

list X
1. behavioural response
2. body core
3. body shell
4. conformer
5. ectotherm
6. effector
7. endotherm
8. hypothalamus
9. internal environment
10. negative feedback
11. physiological homeostasis
12. regulator
13. thermoreceptor
14. thermoregulation
15. vasoconstriction
16. vasodilation

list Y
a) human body's community of cells and the tissue fluid that bathes them
b) maintenance of body's internal environment within tolerable limits by negative feedback control
c) region of the brain containing a centre which regulates body temperature
d) muscle or gland which performs the body's response to stimuli following receipt of signals from the nervous system
e) structure which detects changes in body temperature
f) organism able to control its internal environment and be independent of its external environment
g) organism unable to control its internal environment and dependent on its external environment
h) mechanism of homeostasis whereby a change in a physiological factor triggers a response that counteracts the original change
i) animal which is able to regulate its body temperature by physiological means
j) animal which is unable to regulate its body temperature by physiological means
k) outer layers and limbs of body which have a 'superficial body' temperature of around 33°C
l) vital organs of body normally at 'deep body' temperature of around 37°C
m) process by which the bore of skin arterioles become narrower
n) process by which the bore of skin arterioles become wider
o) mechanism adopted by some conformers (unable to employ physiological mechanisms) to maintain an optimum metabolic rate
p) maintenance of a mammal's internal body temperature within a tolerable range

Multiple choice test
Choose the ONE correct answer to each of the following multiple choice questions.

1 The following table compares conformers and regulators. Which line is NOT correct?

		conformer	regulator
A	state of animal's internal environment in relation to factors affecting its external environment	directly dependent	not directly dependent
B	metabolic costs	high	low
C	range of ecological niches that can be exploited	narrow	wide
D	level of adaptability to environmental change	low	high

Table 12.1

Questions 2 and 3 refer to the following graph which charts the variation in salt concentration of the body fluid of four invertebrate animals when placed in different dilutions of sea water.

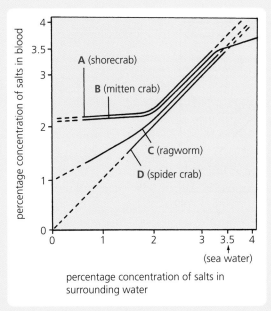

percentage concentration of salts in surrounding water

Figure 12.1

2   Which animal is completely UNABLE to maintain a blood concentration of salt above that in the surrounding water?
3   Which animal is able to maintain a blood concentration of salt lower than that in the surrounding water?
4   Negative feedback control involves the following four stages.
    1 effectors bring about corrective responses
    2 a receptor detects a change in the internal environment
    3 deviation from the norm is counteracted
    4 nerve or hormonal messages are sent to effectors
    The order in which these occur is
    A   2,1,4,3   B   2,4,1,3   C   2,4,3,1   D   4,2,1,3
5   Bacteria taken from the intestine of a certain animal were found to grow on a culture plate at 7°C but not at 37°C. From which animal had they come?
    A   cod   B   seal   C   penguin   D   polar bear

6 Which of the following graphs best shows the relationship between external temperature and body temperature of an endotherm and an ectotherm?

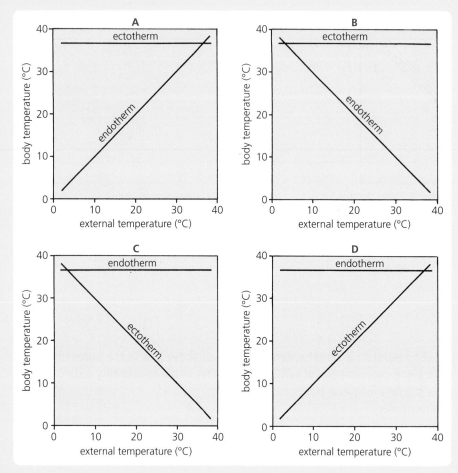

Figure 12.2

7 The responses to external environment shown in the following table are typical of one of the animals given below the table. Which one?

temperature of habitat (°C)	activity of animal	metabolic rate
0	inactive	low
10	inactive	low
20	active	normal
30	very active	high

Table 12.2

A lizard   B gerbil   C vulture   D camel

8 Which of the following animals are BOTH endotherms?
A whale and herring   B herring and shark   C shark and dolphin   D dolphin and whale

9 The temperature-monitoring centre in the human brain is situated in the
A hypothalamus.   B pituitary.   C medulla.   D cerebellum.

10 The following diagram shows a section through the skin of a mammal.

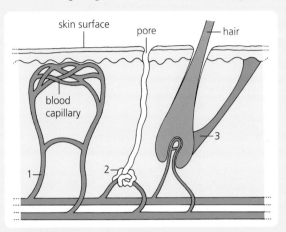

Figure 12.3

Which line in the table below represents the responses of the numbered structures to a drop in temperature of the body core following the ingestion of crushed ice?

	number of structure		
	1	**2**	**3**
A	vasodilation	decreased activity	relaxation
B	vasoconstriction	increased activity	contraction
C	vasoconstriction	decreased activity	contraction
D	vasodilation	increased activity	relaxation

Table 12.3

Questions 11, 12 and 13 refer to the following diagram.

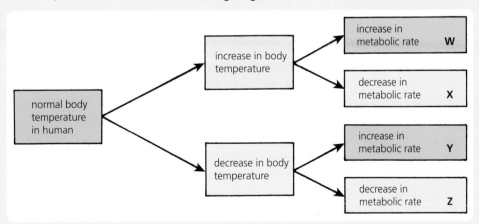

Figure 12.4

11 Which letters indicate the normal negative feedback control of body temperature?

   A W and Y   **B** W and Z   **C** X and Y   **D** X and Z

12 Which situation would be the immediate result of exposure to intense cold?

   A W   **B** X   **C** Y   **D** Z

13 Which situation would be the result of prolonged exposure to intense cold leading to hypothermia?

   A W   **B** X   **C** Y   **D** Z

Questions 14 and 15 refer to the following graph which shows the amount of heat lost by evaporation by a naked person at rest at different external temperatures.

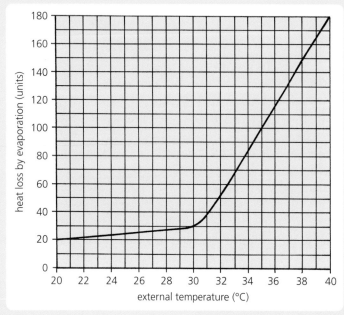

Figure 12.5

14 An increase in heat loss from 60 to 140 units requires an increase in external temperature of

A 4.5°C   B 5.0°C   C 5.5°C   D 6.0°C

15 The percentage increase in heat loss that occurred between external temperatures of 30°C and 38°C was

A 90   B 120   C 400   D 480

13 Metabolism and adverse conditions

Matching test
Match the terms in list X with their descriptions in list Y.

list X
1 aestivation
2 consequential
3 cyclic fluctuation
4 daily torpor
5 dormancy
6 extremophile
7 heat-tolerant DNA polymerase
8 hibernation
9 individual marking
10 innate
11 learned
12 migration
13 predictive
14 transmitter

list Y
a) use of bands or tags to track the route and distance covered by a migratory animal
b) behaviour which is inherited and inflexible
c) behaviour which is gained by experience and is flexible
d) specialised piece of equipment attached to migratory animal which emits signals indicating the animal's route
e) regular variation in environmental conditions often beyond the tolerable limits of an organism's metabolism
f) physiological state in which an animal's metabolic rate becomes reduced for part of each 24-hour cycle
g) period of an organism's life cycle during which growth and development temporarily come to a halt
h) type of dormancy which takes place before the arrival of the adverse conditions
i) type of dormancy which takes place after the arrival of the adverse conditions
j) type of predictive dormancy which enables some animals to survive adverse winter conditions
k) type of dormancy which enables some animals to survive periods of excessive heat and drought
l) enzyme extracted from thermophiles used in the polymerase chain reaction
m) organism that lives in extreme conditions that would be lethal to most other living things
n) regular long distance movement by the members of a species from one place to another where each place offers conditions more favourable than the other for part of the year

Multiple choice test
Choose the ONE correct answer to each of the following multiple choice questions.

1 The diagram below shows four imaginary animals. Which one is best suited to an extremely hot climate?

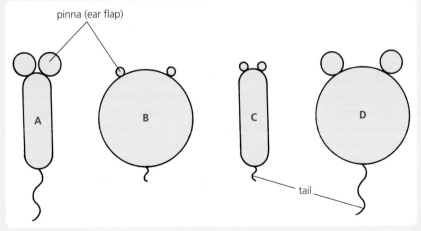

Figure 13.1

2 In autumn, a deciduous tree sheds its leaves and enters a dormant phase. This state is described as

    A predictive and the plant's metabolic rate increases.

    B consequential and the plant's metabolic rate increases.

    C predictive and the plant's metabolic rate decreases.

    D consequential and the plant's metabolic rate decreases.

3 The diagram below shows a section through the skin of a seal pup.

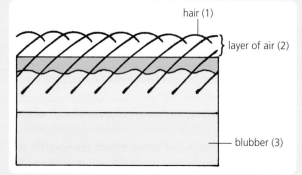

Figure 13.2

Which line in the table below matches each numbered structure with the type of environment to which it is best suited to act as a layer of insulation?

	type of environment	
	terrestrial	aquatic
A	1	2
B	2	3
C	1	3
D	2	1

Table 13.1

Questions 4, 5 and 6 refer to the diagram on the following page. It shows an experiment set up to investigate the effect of variable factors in the breaking of seed dormancy in two species of apple tree (P and Q). The seeds in the diagram have been exposed to varying conditions of temperature (- 4°C or 4°C) and length of exposure time (1 month or 3 months). They are about to be kept at room temperature for 3 weeks and then the different treatments compared.

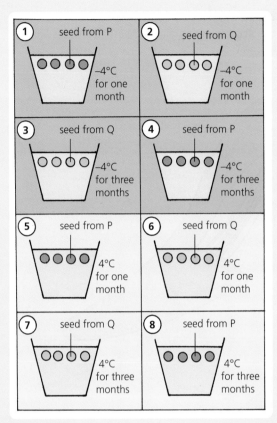

Figure 13.3

4 Which of the following should be compared to find out the effect of temperature on the breaking of dormancy in seeds of species Q?

   A 2 and 3   **B** 2 and 6   **C** 3 and 6   **D** 3 and 8

5 Which of the following should be compared to find out the effect of length of exposure on the breaking of dormancy in seeds of species P?

   A 1 and 5   **B** 4 and 8   **C** 5 and 7   **D** 5 and 8

6 A comparison of set-ups 3 and 8 is invalid because the number of variable factors by which they differ is

   A 1   **B** 2   **C** 3   **D** 4

7 Which line in the following table correctly answers three questions about consequential dormancy?

	Is the organism able to exploit available resources for the maximum time?	Is the organism's activity cut short by its response to decreasing photoperiods?	Does the organism run the risk of death following sudden exposure to adverse conditions?
A	yes	no	yes
B	no	yes	no
C	yes	no	no
D	no	yes	yes

Table 13.2

8 Excessive heat loss from a whale's poorly insulated flipper is prevented by heat being transferred from blood in incoming arteries to blood in returning veins. Which part of the accompanying diagram best illustrates this phenomenon?

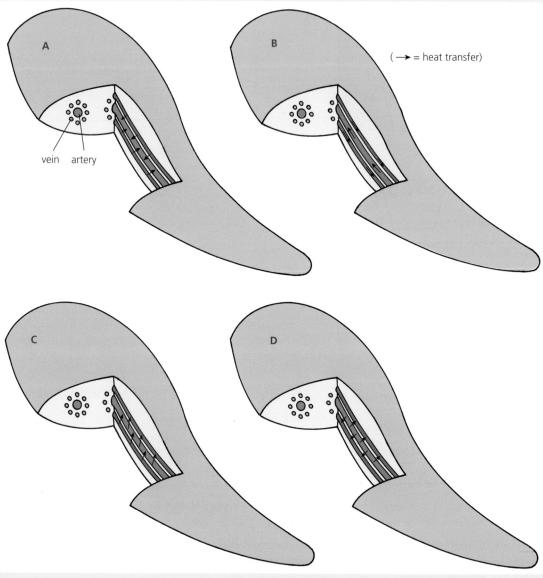

Figure 13.4

9 Which of the following graphs indicates the relationship between body temperature and environmental temperature in a hibernating European hedgehog?

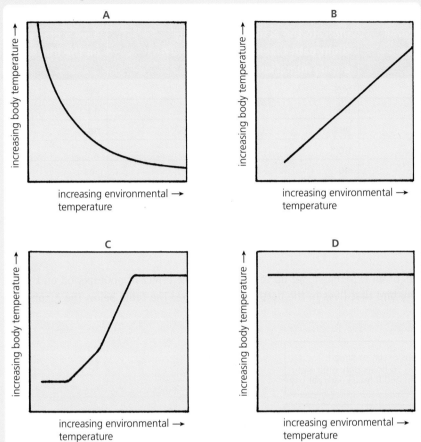

Figure 13.5

10 Aestivation is a form of dormancy that enables an animal to

A survive a period of excessive heat and drought.

C hibernate for part of every twenty-four hour cycle.

B enter a period of daily torpor and conserve energy.

D survive a period of excessive exposure to cold.

11 Three features exhibited by an animal awakening from hibernation are indicated in the following table. Which line is correct?

	metabolic rate	body temperature	heart rate
A	↓	↓	↓
B	↑	↓	↑
C	↓	↑	↑
D	↑	↑	↑

Table 13.3
(Note: ↑ = increase, ↓ = decrease)

12 A hummingbird undergoes a period of torpor every night because

A its metabolic rate is so high that it would become exhausted without a rest.

B its natural habitat is cold both in daytime and at night.

C it feeds on nectar from flowers that close their petals at night.

D its relative surface area is so large that it would lose too much heat.

13 The following table compares two techniques used to study long-distance migration in birds. Which double row correctly answers the four questions posed?

	Technique	Does the bird need to be recaptured for data to be obtained?	Does the technique indicate the actual flight path taken during migration?	Is the technique relatively expensive?	Can the technique have a drag effect on the bird?
A	transmitter	no	yes	yes	yes
	metal band	yes	no	no	no
B	transmitter	yes	yes	yes	no
	metal band	no	no	no	yes
C	transmitter	no	no	no	yes
	metal band	yes	yes	yes	no
D	transmitter	yes	no	yes	yes
	metal band	no	yes	no	no

Table 13.4

14 The following graph shows the results of an experiment set up to investigate the effect of photoperiod on the level of activity of a type of migratory bird that lives in the northern hemisphere. The table below the graph gives details of natural day lengths.

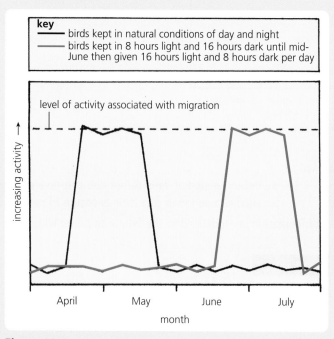

Figure 13.6

time of year	length of day (h)	length of night (h)
mid April	14.0	10.0
mid May	15.5	8.5
mid June	16.5	7.5
mid July	15.5	8.5

Table 13.5

What is the minimum number of hours of light per day needed to trigger the level of activity associated with migration in this type of bird?

A 10.0 B 14.0 C 15.5 D 16.5

15 Innate behaviour is
  A inherited and flexible.   B learned and flexible.
  C inherited and inflexible.   D learned and inflexible.

16 Heat-tolerant DNA polymerase was originally obtained from
  A acidophiles.   B methanogens.
  C thermophiles.   D green sulphur bacteria.

Questions 17, 18 and 19 refer to the diagram below. It shows the results of an experiment carried out on a species of bird that normally migrates in autumn from Britain directly south to West Africa. The birds were captured at point S in Spain and released at point T in Turkey. Some migrated to West Africa and some to East Africa.

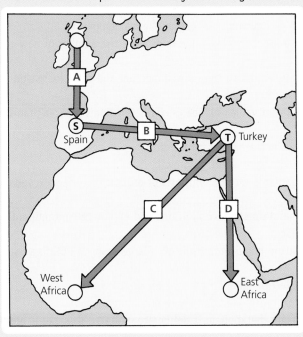

Figure 13.7

17 Which arrow represents the displacement route?
18 Which arrow represents innate behaviour only?
19 Which arrow represents a combination of innate behaviour and learned behaviour?
20 Some extremophiles thrive at temperatures between 60°C and 80°C but do not prosper at 40°C or less. This phenomenon is explained by the fact that
  A their enzymes work best at high temperatures.
  B their metabolic pathways remain unaffected by temperature change.
  C they are able to maintain a cool internal body temperature.
  D their metabolism functions without the use of enzymes.

14 Environmental control of metabolism

Matching test
Match the terms in list X with their descriptions in list Y.

list X
1 aseptic
2 biosynthesis
3 death
4 fermenter
5 mean generation time
6 growth
7 growth medium
8 inducer
9 inhibitor
10 lag
11 log/exponential
12 manipulation
13 precursor
14 primary
15 secondary
16 semi-logarithmic
17 stationary

list Y
a) metabolite which acts on an enzyme at an earlier stage in a pathway exerting negative feedback control
b) metabolite which occurs early in a pathway and gives rise to another metabolite later in the pathway
c) metabolite which ensures that the gene coding for a certain enzyme remains switched on
d) average time taken for a population of cells to double in number
e) phase of growth during which the rate of production of new cells equals the death rate of the old ones
f) phase of growth during which cells grow and multiply at the maximum rate
g) phase of growth during which the number of cells dying exceeds the number of new cells being produced
h) phase of growth during which cells adjust to the growth medium and show increased metabolic activity but no increase in number
i) type of graph paper where each set of values on the y-axis is ten times greater than the set in the previous cycle below
j) metabolite made by a microbe but not essential for its growth
k) metabolite made by a microbe and essential for its growth
l) process that occurs when an organism's rate of synthesis of organic materials exceeds the rate of their breakdown
m) techniques employed to try to maintain sterile conditions and eliminate contaminants while working with micro-organisms
n) specialised container used to culture micro-organisms on a large scale to obtain a useful product
o) building-up of a complex molecule from simpler ones by a cell
p) substance such as nutrient agar or broth used to culture micro-organisms
q) process by which a micro-organism's metabolism is deliberately altered to make it produce large quantities of a metabolite

Multiple choice test
Choose the ONE correct answer to each of the following multiple choice questions.

1 Many micro-organisms are highly valued in research and industrial processes because
   A they reproduce and grow quickly and are easy to cultivate.
   B they produce many useful metabolites without a source of phosphorus in their medium.
   C their structure can be manipulated from unicellular to multicellular in order to boost growth.
   D they are easy to culture and grow without needing an energy source.
2 A standard mass of live yeast was added to glucose solutions at six different pH values. The time taken to collect $5cm^3$ of carbon dioxide for each was recorded as shown in the following table.

pH	time to collect 5cm³ of carbon dioxide (min)
4	340
5	221
6	102
7	119
8	136
9	153

Table 14.1

What percentage decrease in time occurred between pH 4 and pH 7?

A 35   B 60   C 65   D 221

3 Micro-organisms need a source of nitrogen for healthy growth. Which line in the following table is correct?

	possible source of nitrogen	reason why nitrogen is required
A	ammonium compound	as an energy source
B	carbohydrate	as an energy source
C	carbohydrate	for synthesis of nucleic acids
D	ammonium compound	for synthesis of nucleic acids

Table 14.2

4 The following diagram shows four of the steps involved in the streak method of isolating yeast from grape 'juice' containing yeast bloom on a Petri dish of selective medium.

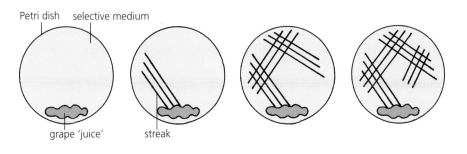

Figure 14.1

Which part of the diagram below shows the missing step in the streaking procedure?

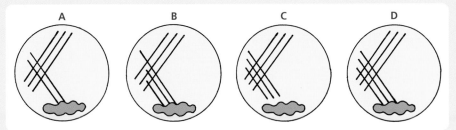

Figure 14.2

Questions 5 and 6 refer to the following bar chart of a country's production of ethanol using biotechnological techniques.

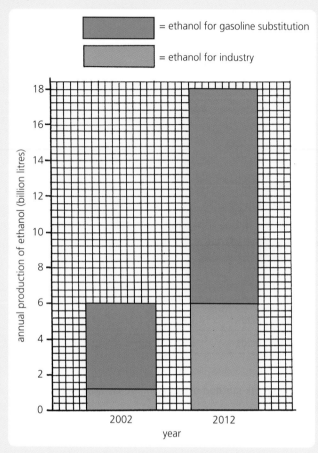

Figure 14.3

5 The production of ethanol for gasoline substitution over the 10-year period increased by a factor of

   A 2.5   B 4.0   C 5.0   D 12.0

6 In 2002 the simple whole number ratio of ethanol for gasoline substitution to ethanol for industry was

   A 2:1   B 3:1   C 4:1   D 5:1

Questions 7, 8 and 9 refer to the following graph which shows the results of culturing the fungus *Penicillium* in a large fermenter to produce the antibiotic penicillin.

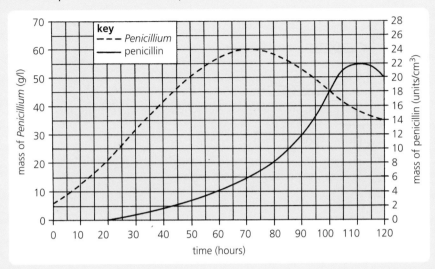

Figure 14.4

7 The mass of penicillin shows the greatest gain between hours
   A 20 and 40.   B 40 and 60.   C 60 and 80.   D 80 and 100.

8 Which line in the following table correctly refers to the mass of fungus and the mass of antibiotic present at 94 hours?

	mass of *Penicillium* (g/l)	mass of penicillin (units/cm³)
A	20	14
B	50	14
C	20	35
D	50	35

Table 14.3

9 The best time to harvest the fungus in order to extract the antibiotic would be at
   A 70 hours.   B 100 hours.   C 110 hours.   D 120 hours.

Questions 10 and 11 refer to the diagram below which shows an industrial fermenter in action.

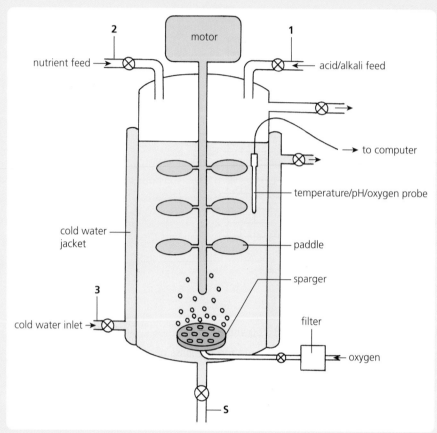

Figure 14.5

10 Sterile materials are introduced into the fermenter at point(s)

  A 1, 2 and 3.   **B** 1 only.   **C** 2 only.   **D** 1 and 2.

11 Structure S is the

  A harvest pipe.   **B** pH controller.   **C** cold water outlet.   **D** exhaust gas outlet.

Questions 12, 13 and 14 refer to the diagram below. It shows a graph of the growth pattern of a type of bacterium.

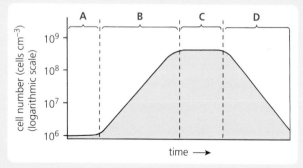

Figure 14.6

12 Which region of the graph represents the lag phase?

13 Which region of the graph represents the phase of growth during which the medium becomes depleted and rate of production of new cells equals rate of death of old ones?

14 Which region of the graph represents the phase of exponential growth?

Questions 15 and 16 refer to the following two tables. The first table shows the results from an experiment where an inoculum of bacteria was added to nutrient medium in a fermenter and given optimum conditions for growth.

time (in 20-minute intervals)	cell number (x 10^3)	cell number (correct to two decimal places)
0	[X]	1.30×10^4
1	26	2.60×10^4
2	52	5.20×10^4
3	104	1.04×10^5
4	208	2.08×10^5
5	416	4.16×10^5
6	832	8.32×10^5
7	1664	[Y]

Table 14.4

	X	Y
A	1.3	1.66×10^6
B	1.3	16.64×10^6
C	13.0	1.66×10^6
D	13.0	16.64×10^6

Table 14.5

15 Which line in the second table gives the correct answers needed to complete boxes [X] and [Y] in the first table?

16 If the bacteria were allowed to continue growing at the same rate, at which 20-minute time interval would the population first exceed 6 million?

A 8   B 9   C 10   D 11

17 Generation time is the time taken for a generation of micro-organisms to divide and double in number.

$q = p \times 2^n$ and $g = \dfrac{t}{n}$

where p = number of bacteria at start

q = number of bacteria after n generations

n = number of generations

g = mean generation time (min)

t = time for n generations (min)

What is the mean generation time (in min) for a colony of bacteria which began as a population of 1000 and had grown to a population of 3.2×10^4 after 200 minutes?

A 10   B 20   C 30   D 40

18 The graph below shows the exponential growth of a type of unicellular micro-organism plotted on semi-logarithmic graph paper.

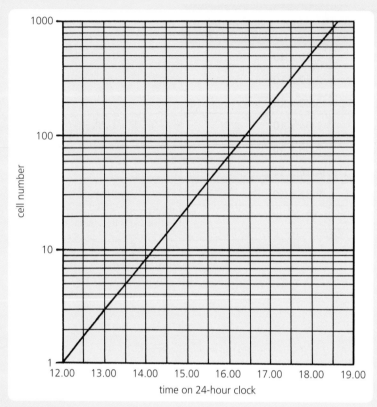

Figure 14.7

By how many times had the population multiplied between 13.00 and 18.30?

A 300	B 450	C 500	D 897

19 A secondary metabolite is produced by a growing filamentous fungus during its

A lag phase and is needed for a later phase of fungal growth.

B stationary phase and may confer an ecological advantage on the fungus.

C lag phase and may confer an ecological advantage on the fungus.

D stationary phase and is needed for a later phase of fungal growth.

20 The diagram below represents a metabolic pathway as it occurs naturally in a certain micro-organism.

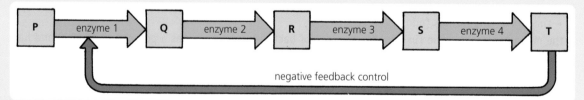

Figure 14.8

In order to mass-produce metabolite R, it may be possible to manipulate this pathway by adding

A metabolite P to act as an inhibitor.

B metabolite Q to act as an inducer.

C metabolite S to act as a precursor.

D metabolite T to exert feedback control.

15 Genetic control of metabolism & 16 Ethical considerations

Matching test
Match the terms in list X with their descriptions in list Y.

list X
1 artificial
2 DNA ligase
3 genetic instability
4 genetic variation
5 marker
6 mutagenesis
7 mutagenic
8 origin of replication
9 plasmid
10 recombinant DNA technology
11 restriction endonuclease
12 restriction site
13 selective breeding
14 sexual
15 sticky end
16 vector

list Y
a) one of two structures formed at a restriction site when DNA is cut by a restriction endonuclease
b) enzyme used to seal a DNA fragment into a bacterial plasmid
c) the location on a plasmid which is cut open by a restriction endonuclease
d) process by which only the organisms with desirable features are chosen as the parents of the next generation
e) process by which genetic material from one organism is inserted into the genome of another
f) enzyme used to cut DNA into fragments and to cleave open bacterial plasmids that are to receive the DNA
g) agent such as a plasmid by means of which a fragment of 'foreign' DNA is inserted into a host cell
h) type of chromosome constructed by scientists to act as a vector in DNA technology
i) form of reproduction by which new genotypes can be produced by fungi but not bacteria
j) tendency of a mutant form of a gene to undergo reverse mutation and revert to the wild-type form
k) heritable diversity that exists among living things
l) the creation of mutations
m) the site on a plasmid which contains genes controlling self-replication of plasmid DNA
n) gene present in a plasmid which enables scientists to determine whether or not a host cell has taken up the plasmid
o) small ring of DNA present in bacteria that is used as a vector in recombinant DNA technology
p) referring to an agent (physical or chemical) that acts on DNA and causes mutations

Multiple choice test
Choose the ONE correct answer to each of the following multiple choice questions.

1 Which of the following statements about micro-organisms is correct?
   A A mutation only occurs in response to exposure to radiation.
   B A mutant allele normally confers an advantage on the organism affected.
   C A mutant strain is unstable and may undergo a reverse mutation.
   D A mutation always leads to an adverse change in an organism's phenotype.

2 The mutation frequency of a bacterial gene can be expressed as the number of mutations that occur at that genetic site per million cells. In *Escherichia coli* it is estimated that the gene for resistance to a type of virus arises spontaneously in 3 out of 10^8 cells. Expressed as a gene frequency per million cells, this would be
   A 0.03   B 0.3   C 30   D 300 ➡

3 Which line in the following table is correct?

	mutagenic agent	effect
A	deep freezing	increased rate of spontaneous mutation
B	UV light	increased rate of spontaneous mutation
C	deep freezing	increased rate of induced mutation
D	UV light	increased rate of induced mutation

Table 15.1

4 The process of site-specific mutagenesis of a gene involves the following procedural steps.
   1 mutant gene introduced back into the cell
   2 many copies of the original gene's DNA produced by PCR
   3 mutated cell cultured to allow study of phenotypic change
   4 base pair sequence at a specific site in the DNA chain altered
   The sequence in which these steps are carried out is
   A 2, 4, 1, 3   **B** 2, 4, 3, 1   **C** 4, 2, 1, 3   **D** 4, 2, 3, 1

5 The diagram below shows the serial dilution of a colony of yeast cells about to be carried out.

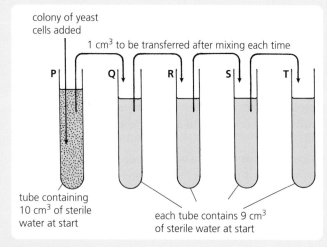

Figure 15.1

Once the procedure has been completed, by how many times will the concentration of yeast cells in tube Q exceed that in tube T?
   A 10^2 **B** 10^3 **C** 10^4 **D** 10^5

Questions 6 and 7 refer to the diagram below which shows the results of an investigation using UV-sensitive yeast to test the effectiveness of sun barrier creams.

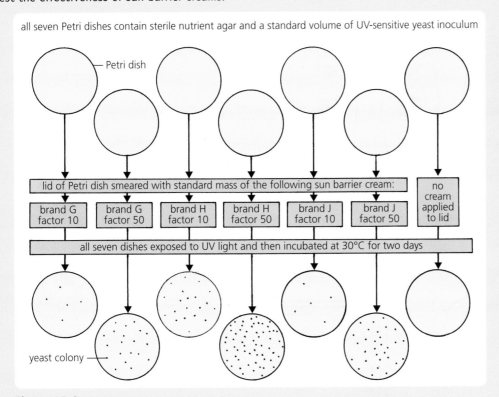

all seven Petri dishes contain sterile nutrient agar and a standard volume of UV-sensitive yeast inoculum

— Petri dish

lid of Petri dish smeared with standard mass of the following sun barrier cream:

| brand G factor 10 | brand G factor 50 | brand H factor 10 | brand H factor 50 | brand J factor 10 | brand J factor 50 | no cream applied to lid |

all seven dishes exposed to UV light and then incubated at 30°C for two days

yeast colony —

Figure 15.2

6   From the results it can be concluded that
   A   brand G offers more protection against UV light than brand H at factor 50.
   B   brand G offers more protection against UV light than brand J at factor 10.
   C   brands H and J offer equal protection against UV light at factor 50.
   D   brands G and J offer less protection against UV light than brand H at factor 10.
7   A further control that could have been included is a Petri dish containing
   A   plain agar and UV-sensitive yeast, given UV light.
   B   plain agar and wild-type yeast, kept in darkness.
   C   nutrient agar and UV-sensitive yeast, kept in darkness.
   D   nutrient agar and wild-type yeast, given UV light.
8   The following diagram represents three strains of a type of micro-organism involved in an improvement breeding plan.

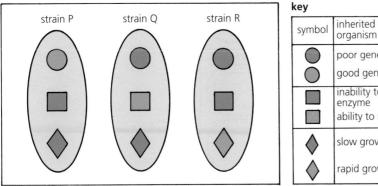

strain P	strain Q	strain R		key

symbol	inherited trait of micro-organism
● (large, dark)	poor genetic stability
● (small)	good genetic stability
■ (dark)	inability to make key enzyme
■ (grey)	ability to make key enzyme
◆ (dark)	slow growth
◆ (grey)	rapid growth

Figure 15.3

The diagram below shows the four types of offspring that result from a cross between P and Q. Which of these offspring should be crossed with strain R to try to produce the best possible strain?

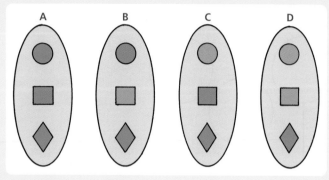

Figure 15.4

9 The left side of the diagram below shows the process of conjugation between two bacteria with different genomes. Which of the boxes on the right side gives the correct information to complete the blank box?

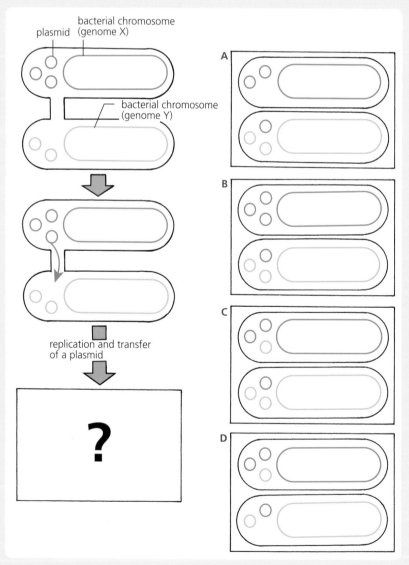

Figure 15.5

Questions 10, 11 and 12 refer to the following information.

A certain strain of bacterium *Streptomyces* possesses a gene that makes it resistant to a type of weedkiller. In order to produce a strain of potato also resistant to the weedkiller, genetic engineers have transferred the useful gene using as the vector a plasmid from a second type of bacterium that induces tumours in potatoes.

The following list gives the steps adopted in the procedure in the WRONG order.

1 gene resistant to weedkiller inserted and sealed into plasmid
2 gene resistant to weedkiller cut out of *Streptomyces* bacterium
3 samples of infected potato cells grown into 'transformed' plants resistant to weedkiller
4 plasmid extracted from tumour-inducing bacterium and cut open
5 plasmid returned to tumour-inducing bacterium which is allowed to infect potato cells

10 Which answer indicates the correct sequence of events?

    A 2, 1, 5, 4, 3   **B** 4, 1, 2, 3, 5   **C** 2, 4, 1, 5, 3   **D** 4, 5, 2, 3, 1

11 The enzyme ligase would be employed at

    A step 1 only.   **B** steps 1 and 2.   **C** step 2 only.   **D** steps 2 and 4.

12 Restriction endonuclease would be employed at steps

    A 1 and 2.   **B** 1 and 4.   **C** 2 and 4.   **D** 2 and 5.

13 Which of the following statements is NOT correct?

    DNA technology is used to insert into a micro-organism one or more genes that

    A cause the cells to secrete the useful product into the surrounding medium.
    B alter the specific metabolic steps thereby amplifying manufacture of the product.
    C remove the inhibitory controls that would normally regulate a pathway.
    D enable the micro-organism to survive in the external environment.

Questions 14 and 15 refer to the following information.

Restriction endonuclease enzymes do not cut DNA at random but recognise particular sequences of bases.

14 One enzyme has the following recognition sequence:

```
        cut
         ↓
- - - -G G A T C C- - - -
- - - -C C T A G G- - - -
         ↑
        cut
```

The diagram below shows a piece of DNA about to be cut.

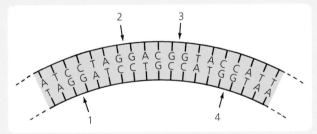

Figure 15.6

At which pair of numbered sites would the enzyme make its cuts?

    A 1 and 2   **B** 3 and 4   **C** 1 and 3   **D** 2 and 4   ➡

15 The following diagram shows a different piece of DNA about to be acted upon by a second enzyme with the recognition sequence:

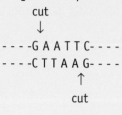

Figure 15.7

Which of the diagrams below shows the outcome of this enzyme's action?

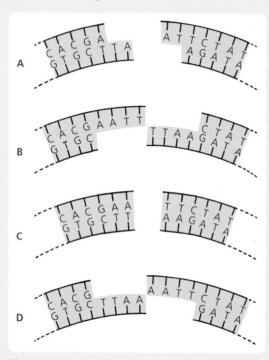

Figure 15.8

16 The following diagram shows a plasmid about to be used as a vector.

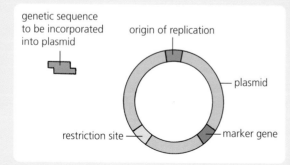

Figure 15.9

Which letter in the diagram below indicates the same plasmid in use?

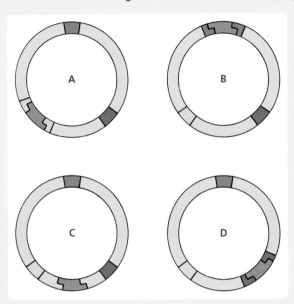

Figure 15.10

17 The following diagram shows four DNA sequences. Which of these could be recognised as a cutting site by a restriction endonuclease that creates sticky ends?

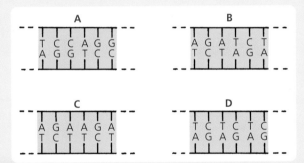

Figure 15.11

18 On some occasions an artificial chromosome is a more effective vector than a plasmid because only the artificial chromosome
   A has a restriction site that can be opened by a restriction endonuclease.
   B is able to carry a long sequence of foreign DNA to the recipient cell.
   C bears a marker gene that can be used by scientists in the selection process.
   D contains genes for self-replication and regulatory sequences.

19 The use of a bacterium as the host cell to a genetic sequence from a plant or animal is limited by the fact that the
   A protein formed carries unnecessary post-translational modifications.
   B pre-transcriptional splicing of the mRNA transcript fails to occur.
   C polypeptide formed may be incorrectly folded and be inactive.
   D insertion of a gene from a prokaryote into a bacterium is problematic.

20 Which area in the following Venn diagram represents the successful discovery and development of a new microbiological product?

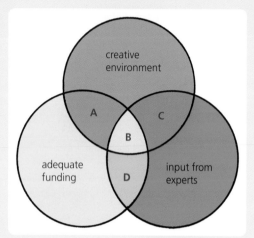

Figure 15.12

Unit 3

Sustainability and Interdependence

17 Food supply, plant growth and productivity

Matching test part 1

Match the terms in list X with their descriptions in list Y.

list X	list Y
1 absorption	a) yellow pigments which absorb light and pass the energy to chlorophyll
2 absorption spectrum	b) ability to access food of adequate quantity and quality
3 action spectrum	c) variety of plant obtained from natural species by selection and/or genetic manipulation and maintained by cultivation
4 carbohydrates	
5 carotenoids	d) technique used to separate the components of a mixture that differ in their degree of solubility in a solvent
6 chlorophylls	
7 chromatography	e) process by which green plants trap light energy and use it to produce carbohydrates
8 cultivar	f) process by which light is taken in and retained by a leaf
9 food security	g) process by which light passes through a leaf
10 nanometre	h) process by which light bounces off a leaf surface
11 photosynthesis	i) graph showing the rate of photosynthesis by a green plant at different wavelengths of light
12 reflection	j) display/graph showing the quantity of light absorbed by a pigment at different wavelengths of light
13 transmission	
14 trophic	k) term referring to a level in a food chain or pyramid
	l) green pigments which absorb light in the red and blue regions of the spectrum
	m) large group of organic compounds that contain carbon, hydrogen and oxygen
	n) one thousand-millionth of a metre

Matching test part 2

Match the terms in list X with their descriptions in list Y.

list X	list Y
1 ATP	a) metabolite in the Calvin cycle which acts as the carbon dioxide acceptor
2 ATP synthase	b) breakdown of water during the light-dependent stage of photosynthesis
3 Calvin cycle	c) raw material which becomes split into oxygen and hydrogen during photolysis
4 carbon dioxide	d) compound which accepts hydrogen released during the photolysis of water
5 electron transport chain	e) second stage of photosynthesis during which carbon dioxide is taken up and sugar is synthesised
6 glyceraldehyde-3-phosphate (G3P)	f) enzyme which catalyses the synthesis of ATP from ADP and P_i
7 hydrogen	g) enzyme which fixes carbon dioxide by attaching it to RuBP
8 light-dependent reaction	h) high energy compound which provides the energy needed to drive the Calvin cycle
9 NADP	i) product of photolysis of water which is required for aerobic respiration
10 oxygen	j) product of photolysis of water which becomes attached to NADP
11 photolysis	k) first stage of photosynthesis during which light energy is converted to chemical energy
12 ribulose bisphosphate (RuBP)	l) metabolite in the Calvin cycle used to regenerate RuBP and make sugar
13 rubisco (RuBisCO)	m) raw material which supplies carbon atoms to be fixed into carbohydrate
14 water	n) group of protein molecules in the membrane of a chloroplast which makes energy available to pump H^+ across the membrane and to split water molecules

Matching test part 3

Match the terms in list X with their descriptions in list Y.

list X

1 assimilation
2 biological yield
3 biomass
4 economic yield
5 harvest index
6 leaf area index
7 limiting
8 net assimilation
9 productivity

list Y

a) measure calculated by dividing dry mass of economic yield by dry mass of biological yield
b) mass of the desired product such as barley grains
c) factor whose restricted supply prevents an increase in the rate of a process
d) process by which food produced by photosynthesis is converted into complex constituents of plant cell systems
e) total biomass of plant material produced
f) rate of generation of new biomass per unit area per unit of time
g) ratio of total leaf area to area of ground covered by the leaves on plants
h) total mass of organic material present in one or more living organisms
i) measure of increase in a plant's biomass due to photosynthesis minus loss due to respiration

Multiple choice test

Choose the ONE correct answer to each of the following multiple choice questions.

1 Which of the following does NOT refer directly to a community's food security?
A The food supply is grown locally by traditional methods.
B A sufficient quantity of food is available to feed everyone.
C The people have the economic means to buy available food.
D The food provides the people with a healthy, balanced diet.

Questions 2, 3 and 4 refer to the following bar chart which shows eight crop plants which provided the world with much of its food during one year.

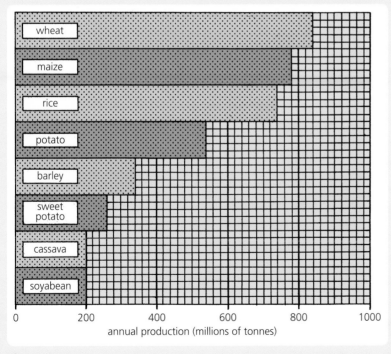

Figure 17.1

2 What was the TOTAL production of food (in millions of tonnes) obtained that year from these plant species?

    A 820   **B** 840   **C** 3900   **D** 3950

3 The total production of food for all plant species that year was 5200 million tonnes. What percentage of the world's total food was derived from sweet potato?

    A 4.42   **B** 5.00   **C** 6.67   **D** 10.38

4 Which plant ALONE provided the world with 15% of its needs?

    A wheat   **B** maize   **C** rice   **D** potato

5 Which of the following procedures could make the most significant contribution to global food security?

    A Incorporation of genetic variety from domestic strains of farm animals into their wild-type relatives.

    B Development and increased use of non biodegradable pesticides and herbicides.

    C Use of DNA recombinant technology to develop crop varieties resistant to insects.

    D Establishment of agricultural zones in desert regions warmed by climate change.

6 In the following diagram of sunlight striking a green leaf, which arrow represents light being transmitted by the leaf?

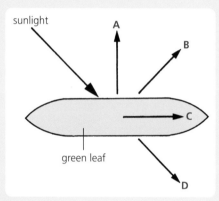

Figure 17.2

7 Which of the following diagrams best represents the absorption spectrum that results when a chlorophyll extract is placed in a beam of white light?

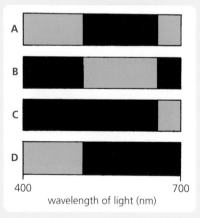

Figure 17.3

8 The following diagram shows an experiment set up to separate photosynthetic pigments by paper chromatography.

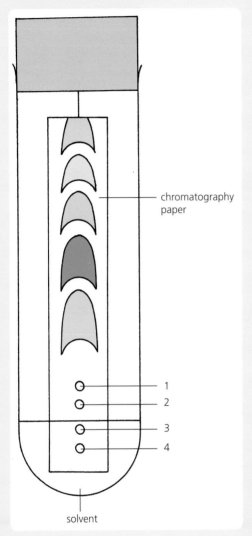

chromatography paper

1
2
3
4

solvent

Figure 17.4

Two alternative positions at which the chlorophyll extract could have been spotted to give this separation are

A 1 and 2.   B 1 and 3.   C 2 and 3.   D 3 and 4.

9 Which of the following graphs best represents the absorption spectrum of BOTH green chlorophyll (Ch) and the yellow carotenoid pigments (Y)?

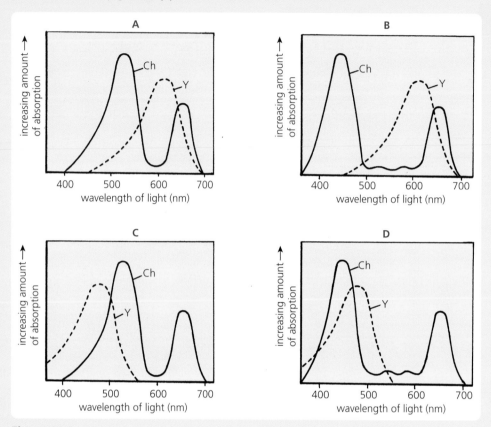

Figure 17.5

10 The diagram below shows the result of an experiment in which a strand of alga was placed in water containing motile aerobic bacteria and illuminated by a microspectrum of white light.

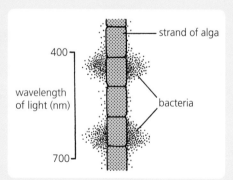

Figure 17.6

Which of the following correctly explains the distribution of bacteria?
A The bacteria absorb red and blue light.
B The alga receives carbon dioxide at these positions.
C The bacteria receive oxygen at these positions.
D The alga feeds on the bacteria.

11 Which of the following statements about carotenoid pigments is correct?

A They limit the range of wavelengths of light absorbed and compete with chlorophyll.

B They extend the range of wavelengths of light absorbed and pass the energy on to the chlorophyll.

C They limit the range of wavelengths of light absorbed enabling the energy to bypass chlorophyll.

D They extend the range of wavelengths of light absorbed enabling them to be independent of chlorophyll.

Questions 12 and 13 refer to the information in the four boxes below which refer to the energy capture and transfer at the start of photosynthesis.

(1) excited electrons are captured by primary electron acceptors

(2) energy is used to drive two biochemical reactions

(3) light is absorbed by pigment molecules

(4) high energy electrons are transferred through electron transport chains

Figure 17.7

12 The correct order in which these processes occur is

  A 1, 3, 2, 4   **B** 3, 1, 2, 4   **C** 1, 3, 4, 2   **D** 3, 1, 4, 2

13 The two biochemical reactions that the energy is used to drive are

A ATP generation and photolysis.

B photolysis and the Calvin cycle.

C the Calvin cycle and the return flow of hydrogen ions.

D the return flow of hydrogen ions and ATP generation.

14 The apparatus shown in the following diagram was set up to investigate photolysis. DCPIP (dichlorophenol-indophenol) is a chemical that acts as a hydrogen acceptor by undergoing the following chemical reaction:

$$DCPIP + 2H^+ + 2e^- \rightarrow DCPIPH_2$$
(dark blue)   (colourless)

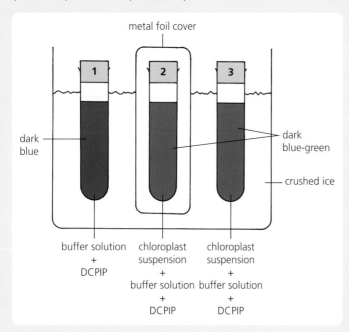

Figure 17.8

Which line in the following table correctly represents the results after 30 minutes in bright light?

	tube 1	tube 2	tube 3
A	dark blue	dark blue-green	green
B	colourless	green	dark blue
C	dark blue	dark blue-green	colourless
D	colourless	dark blue	green

Table 17.1

15 The hydrogen released from water during the light-dependent stage of photosynthesis is transferred to

A NADP.

B chlorophyll.

C ATP synthase.

D ribulose bisphosphate.

Questions 16, 17 and 18 refer to the following diagram of the Calvin cycle.

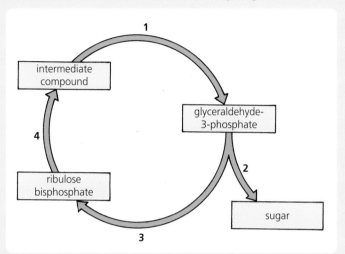

Figure 17.9

16 Carbon dioxide is taken into the cycle at stage
   A 1   B 2   C 3   D 4

17 Energy from ATP is used to drive stages
   A 1 and 2.   B 1 and 3.   C 2 and 3.   D 3 and 4.

18 Rubisco plays its role at stage
   A 1   B 2   C 3   D 4

19 The productivity of an ecosystem was found to be 2.6 grams per square metre per year. An abbreviated version of this information would be written symbolically as
   A $2.6 \, g \, m^2 \, y^{-1}$ B $2.6 \, g^{-1} \, m^2 \, y$ C $2.6 \, g \, m^{-2} \, y$ D $2.6 \, g \, m^{-2} \, y^{-1}$

Questions 20 and 21 refer to the following graph which shows the effect of carbon dioxide concentration on the rate of photosynthesis at three different intensities of light.

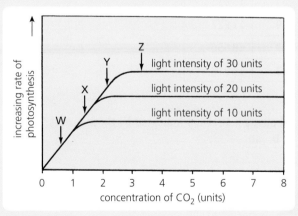

Figure 17.10

20 At which of the following concentrations was carbon dioxide always the limiting factor?
   A 0–1 units   B 1–2 units   C 2–3 units   D 3–4 units

21 Light intensity was the limiting factor at
   A point W only.   B point Z only.   C points X and Y.   D points W, X, Y and Z.

22 The following graph shows the relationship between net productivity and leaf area index for a certain crop plant.

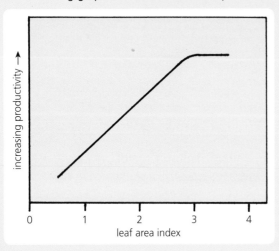

Figure 17.11

The optimum level occurs when
A ground area covered by the leaves is three times that of the total leaf area.
B ground area covered by the leaves is four times that of the total leaf area.
C total leaf area is three times that of the ground area covered by the leaves.
D total leaf area is four times that of the ground area covered by the leaves.

Questions 23 and 24 refer to the following table. It shows the effect of increasing crop planting density on productivity of a type of cereal plant.

crop planting density (no. of plants x 10^3 ha^{-1})	dry biomass at harvest (kg ha^{-1})			
	farm A	farm B	farm C	farm D
20	11 522	12 379	11 530	12 654
25	12 377	14 286	12 261	15 331
30	13 723	15 594	13 949	17 227
35	14 281	16 727	13 887	18 656
40	11 817	14 566	11 623	16 918

Table 17.2

23 What is the optimum level of crop planting density (in no. of plants x 10^3 ha^{-1}) indicated by these results?
   A 25   B 30   C 35   D 40

24 Which farm has the poorest overall productivity?
25 The following table refers to the mean values for a farm's maize crop.

part of maize plant	dry mass per plant (g)
roots	48
leaves	150
stems	192
grains	210

Table 17.3

This farm's harvest index (as a percentage) for maize is
A 8   B 25   C 32   D 35

18 Plant and animal breeding

Matching test
Match the terms in list X with their descriptions in list Y.

list X
1 cultivar
2 deleterious
3 genetic transformation
4 genome sequencing
5 hybrid vigour
6 hybridisation
7 inbreeding
8 inbreeding depression
9 outbreeding
10 randomisation
11 replicate
12 selective breeding
13 test cross
14 treatment

list Y
a) process involving the mating of unrelated members of the same species
b) deterioration in the quality of a strain of natural outbreeders that have been forcibly inbred
c) process by which one variety of organism is crossed with a different variety to try to produce offspring better than either parent
d) improved condition shown by offspring of cross between two different inbred parental strains
e) referring to alleles of genes whose expression results in the formation of harmful characteristics
f) process by which close relatives are bred with one another and prevented from breeding at random
g) process by which only the organisms with the best features are chosen as the parents of the next generation
h) alteration of an organism's genome by the insertion of a DNA sequence from a different organism
i) procedure carried out between an organism whose genotype for a trait is unknown and an organism who is homozygous recessive for the trait
j) variety of plant obtained from natural species by selection and/or genetic manipulation and maintained by cultivation
k) establishing the order of the nucleotide bases all the way along an organism's DNA
l) way in which a plot is dealt with compared with other plots
m) disordered arrangement of replicate treatments to eliminate bias
n) one of several copies of a treatment set up to reduce the effect of experimental error

Multiple choice test
Choose the ONE correct answer to each of the following multiple choice questions.

1 Which of the following is NOT an example of a crop improved by manipulation of its heredity?
   A tomato plants that show increased resistance to eelworm
   B maize plants that are able to thrive in cold, damp conditions
   C soya plants with beans that contain increased protein content
   D potato plants with tubers that show increased susceptibility to blight

2 In a certain breed of cattle, black coat colour (allele B) is dominant to brown coat colour (allele b). The offspring of a cross between a black bull (BB) and a brown cow were allowed to interbreed. What percentage of the progeny would have black coats on average?
   A 25   B 50   C 75   D 100

➡

3   In maize plants, two alleles of the gene for seed colour exist. Purple (P) is dominant to yellow (p). A testcross was carried out to determine the genotype of a certain purple plant. Which line in the following table is correct?

	phenotypic ratio of offspring resulting from backcross	genotype of purple parent
A	1 purple : 1 yellow	heterozygous
B	3 purple : 1 yellow	homozygous
C	1 purple : 1 yellow	homozygous
D	all purple	heterozygous

Table 18.1

Questions 4 and 5 refer to the following information. The diagram below shows a field trial of plots set up to investigate the effect of a fungicide on a new cultivar of barley crop.

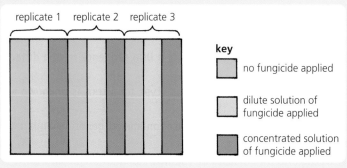

key

□ no fungicide applied

□ dilute solution of fungicide applied

■ concentrated solution of fungicide applied

Figure 18.1

4   Which part of the following diagram shows an arrangement of the plots set up to eliminate bias?

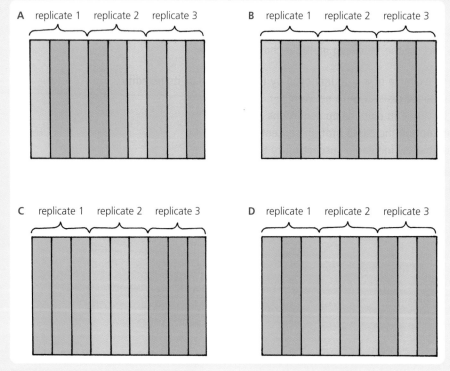

Figure 18.2

5 Several replicates were included in this type of field trial in order to
   A randomise the treatments.
   B minimise experimental error.
   C treat the plots in an orderly sequence.
   D allow a fair comparison between two variable factors.

6 The following table compares features relating to natural inbreeders and natural outbreeders. Which line is NOT correct?

	natural outbreeders	natural inbreeders
A	exemplified by cross-pollinating plants	exemplified by self-pollinating plants
B	homozygosity promoted	heterozygosity promoted
C	cross-fertilisation employed	self-fertilisation employed
D	recessive deleterious alleles masked by dominant alleles	recessive deleterious alleles eliminated by natural selection

Table 18.2

Questions 7 and 8 refer to the following diagram which charts the effect of repeated self-pollination on heterozygosity in a variety of flowering plant.

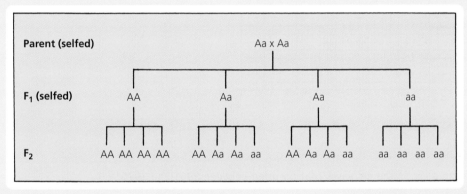

Figure 18.3

7 What percentage of the F_2 generation are heterozygous?
   A 4   B 12   C 25   D 50

8 If the pattern of selfing were repeated, by which generation would there be less than 4% of heterozygotes remaining in the population?
   A F_3 B F_4 C F_5 D F_6

9 The following statements refer to the enforced inbreeding of organisms that are natural outbreeders. Which one is FALSE?

   A Inbreeding increases the chance of individuals arising that are double recessive for an inferior allele.

   B Inbreeding depression often results from hybridisation between unrelated species.

   C Inbreeding results in loss of genetic diversity among members of a domesticated variety.

   D Inbreeding promotes the retention of desirable characteristics in a variety from generation to generation.

Questions 10 and 11 refer to the following graph. It shows the effect of inbreeding and selection on the percentage butterfat and milk yield obtained from a breed of cattle over a 40-year period.

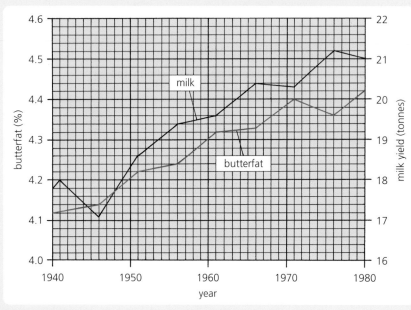

Figure 18.4

10 Which line in the following table shows the increases that occurred between 1951 and 1976?

	butterfat (%)	milk yield (tonnes)
A	0.14	0.24
B	0.14	2.60
C	0.17	0.24
D	0.17	2.60

Table 18.3

11 What percentage decrease in milk yield occurred between 1941 and 1946?

   A 0.09   **B** 0.90   **C** 2.14   **D** 5.00

12 An inbred high-yielding strain of wheat is susceptible to root rot caused by a fungus. A mutant variety of wheat is found to be resistant to the fungus. This useful feature could be introduced to the inbred wheat by cross-breeding it with the mutant variety and then carrying out a series of backcrosses to the

   A inbred line to dilute unwanted genetic material from the mutant.

   B mutant line to concentrate the useful genetic material from the inbred line.

   C inbred line to concentrate the useful genetic material from the mutant.

   D mutant line to dilute the unwanted genetic material from the inbred line.

13 Which of the following statements is FALSE?

   A A hybrid is the result of a cross between genetically dissimilar parents.

   B A hybrid is often stronger or better in some way than its parents.

   C A hybrid formed from two different species is sterile because its chromosomes fail to pair at gamete formation.

   D A hybrid tends to be homozygous for many genes as a result of many generations of inbreeding.

14 The first table below shows the outcome of selfing four breeds of cattle (Q, R, S and T). The second table shows the outcome of hybridisation crosses involving the four breeds of cattle.

parents	average live weight of offspring at 18 months (kg)
Q x Q	300
R x R	350
S x S	250
T x T	300

Table 18.4

parents	average live weight of offspring at 18 months (kg)
Q x R	320
R x S	310
Q x S	280
S x T	290

Table 18.5

Which of the following crosses fails to show hybrid vigour since the offspring are poorer than the mean of the two parents?

   A Q x R   **B** R x S   **C** Q x S   **D** S x T

15 P_1 in the cross shown below represents a cultivated variety of plant that contains the genes for many desirable characteristics. However, P_1 is susceptible to a particular virus. P_2 is a wild variety of the same species that possesses the gene for resistance to the virus.

Following the first cross, 50% of the genetic material (including the resistance gene) received by P_3 comes from the wild parent. In order to dilute this unwanted wild genetic contribution but retain the resistance gene, a series of crosses against P_1 is carried out. The first is shown in the diagram and results in the formation of P_4.

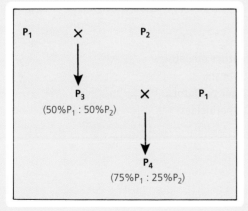

Figure 18.5

How many more crosses will need to be carried out to reduce the wild genetic material inherited by offspring to less than 5%?

   A 1   **B** 2   **C** 3   **D** 4

16 Pedigree dogs are produced by mating closely related members of the same breed with one another. This often results in the production of offspring suffering conditions which affect their fitness. For example, Bulldogs have problems with their breathing and Labradors are prone to arthritis. This phenomenon is known as

A inbreeding depression.

B natural selection.

C hybrid depression.

D heterozygote selection.

17 In order to produce a supply of hybrids showing genetic uniformity, horticulturists often maintain two different true-breeding parental lines of bedding plant. The hybrids cannot be used as the parents of the next generation because

A a high mutation rate occurs among hybrid gametes.

B hybrids of annual plants always form sterile seeds.

C hybrid vigour cannot be passed on to the next generation.

D the hybrids are heterozygous and therefore not true-breeding.

Questions 18 and 19 refer to the following table which shows the results of a study of calving difficulty in four breeds of cattle.

age of cow (years)	percentage calving difficulty			
	breed 1	breed 2	breed 3	breed 4
2	53	30	46	39
3	16	14	15	11
4	7	5	6	5
5	5	3	4	3

Table 18.6

18 It can be correctly concluded from the table that

A as age of cow increases, percentage calving difficulty increases.

B percentage calving difficulty is higher in breed 2 than breed 3.

C among two-year-old cows, members of breed 2 have more difficulty calving than those of breed 4.

D three-year-old cows of breed 4 have less difficulty calving than those of breed 1.

19 The mean percentage calving difficulty for two-year-old cows among these four breeds is

A 13   B 14   C 42   D 52

20 In recent times, scientists have produced crops that contain a gene from a soil bacterium which codes for a protein toxic to certain insects.

This is an example of

A biological control that has made insect pests more susceptible to other predators.

B genetic transformation that has produced plant breeds resistant to pests.

C biological control that has made insect pests less susceptible to their predators.

D genetic transformation that has produced plant breeds resistant to herbicides.

19 Crop protection

Matching test

Match the terms in list X with their descriptions in list Y.

list X
1 annual
2 biological control
3 chemical
4 contact
5 crop rotation
6 cultural
7 integrated pest management
8 monoculture
9 pathogen
10 perennial
11 persistent
12 pesticide
13 resistant
14 selective
15 systemic
16 weed

list Y
a) chemical used to control pests
b) type of non-selective herbicide that destroys all green plant tissue which it meets
c) type of herbicide that mimics plant growth substances and kills broad-leaved plants by stimulating their growth to a harmful extent
d) type of herbicide that is absorbed by a plant and transported internally to all parts where it has a lethal effect
e) disease-causing micro-organism
f) weed that continues to grow for several years
g) weed that completes its life cycle within one year
h) means of control of weeds, pests and pathogens using herbicides, pesticides and fungicides
i) traditional non-chemical means of controlling weeds, pests and pathogens that affect crop plants
j) agricultural practice whereby each of a series of different crop plants is grown in turn on the same piece of ground
k) combination of techniques including chemical, biological and cultural means of control
l) reduction of a pest population by the deliberate introduction of one of its natural enemies
m) unwanted plant that poses economic problems when it is able to multiply among and compete with a crop
n) vast population of a single species cultivated for economic efficiency
o) type of pest that survives treatment by a pesticide and enjoys a selective advantage
p) non-biodegradable chemical whose molecules accumulate along a food chain

Multiple choice test

Choose the ONE correct answer to each of the following multiple choice questions.

1 The following table compares the characteristics of two types of ecosystem. Which line is NOT correct?

	characteristic	natural ecosystem	agricultural ecosystem
A	state of community structure	simpler	more complex
B	total number of different species present	larger	smaller
C	level of genetic diversity present among producers	higher	lower
D	growing conditions available to weeds and pests	more limited	more extensive

Table 19.1

2 Which of the following are BOTH competitive adaptations commonly shown by perennial weeds but normally absent from annual weeds?
A rapid growth and short life cycle
B short life cycle and ability to reproduce vegetatively
C ability to reproduce vegetatively and to produce storage organs
D production of storage organs and a vast number of seeds

➡

3 Which combination of characteristics in the following table best describes cultural means of controlling pests that affect crops?

	preventative	curative	traditional	recently developed	chemical	non-chemical
A	X	✓	X	✓	X	✓
B	✓	X	✓	X	✓	X
C	X	✓	X	✓	✓	X
D	✓	X	✓	X	X	✓

Table 19.2

(Note: ✓ = yes, X = no)

4 Which of the following statements is INACCURATE?
Crop rotation
   A helps to maintain soil fertility.
   B plays a role in breaking pest cycles.
   C helps to keep crop plants healthy.
   D introduces predators to control pests.

5 The following diagram shows the action of a herbicide on a broad-leaved weed growing among a cereal crop.

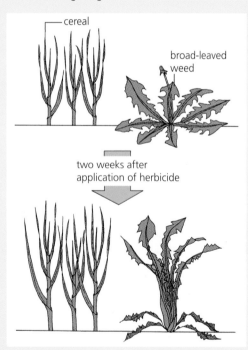

cereal

broad-leaved weed

two weeks after application of herbicide

Figure 19.1

Which line in the table correctly applies to this situation?

	type of herbicide	effect on weed
A	selective	growth stimulated to excessive and harmful extent
B	non-selective	growth stimulated to excessive and harmful extent
C	selective	general destruction of all green plant tissue
D	non-selective	general destruction of all green plant tissue

Table 19.3

Questions 6 and 7 refer to the following bar chart. It represents the results of an investigation into the incidence and viability of potato cyst nematode (PCN) cysts in four soil types.

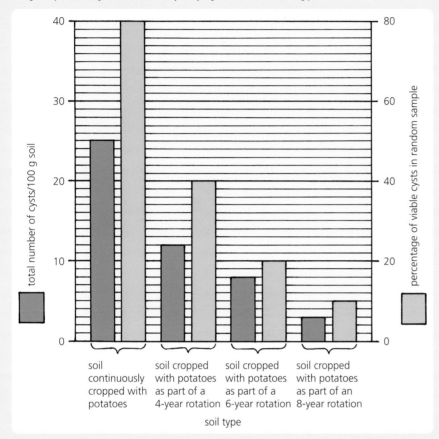

Figure 19.2

6 What percentage decrease in the total number of cysts per 100 g soil occurred between soil continuously cropped with potatoes and soil cropped with potatoes as part of a 6-year rotation?

  A 13   B 17   C 52   D 68

7 Which of the following conclusions can be correctly drawn from the results?

  A Eight years of crop rotation completely solved the problem of PCN cysts in fields cropped with potatoes.

  B The longer the period of crop rotation the more effective the control of PCN cysts in potato fields.

  C The total number of cysts is inversely proportional to the percentage of viable cysts.

  D Continuously cropped soil is four times more likely to contain viable cysts than soil in an 8-year rotation.

Questions 8 and 9 refer to the following graph which shows the effect of the recommended concentration of a selective herbicide on a cereal crop and three different species of broad-leaved weed at different stages of their development.

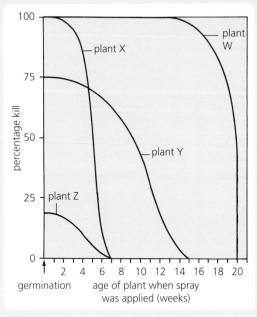

Figure 19.3

8 Which curve represents the cereal crop?

   A W   **B** X   **C** Y   **D** Z

9 In which of the following weeks would application of this concentration of the herbicide spray be MOST effective?

   A 4   **B** 7   **C** 15   **D** 20

10 The following table compares two types of herbicide. Which line is NOT correct?

	type of herbicide	
	systemic	**contact**
A	absorbed by plant and transported internally	not transported internally by plant
B	faster acting	slower acting
C	affects all plant parts	affects green tissues only
D	has lethal effect since no part of plant survives	has non-lethal effect since roots survive allowing regeneration of plant

Table 19.4

Questions 11, 12 and 13 refer to the following table which summarises the results from an investigation into the impact of a new pesticide on four pests.

crop	pest	region of host attacked	average loss of crop (acres/year)	
			without insecticide	with insecticide
apple	aphid	leaf and flower	12 000	600
pea	weevil	leaf and pod	4 500	450
potato	leatherjacket	root and tuber	1 800	1 700
cabbage	caterpillar	leaf and leaf stalk	3 200	400

Table 19.5

11 The number of acres of pea crop saved per year by the use of the chemical was
  A 450   B 4005   C 4050   D 4500

12 On which crop did the chemical have the GREATEST effect relative to the others?
  A apple   B pea   C potato   D cabbage

13 On which pest was the insecticide LEAST effective?
  A leatherjacket   B weevil   C aphid   D caterpillar

Questions 14 and 15 refer to the following table. It records the results of an investigation into the effect of fungicide on four varieties of apple.

variety of apple	fungicide absent (–) or present (+)	fruit lost due to rotting by fungi (%)		
		March	May	July
A	–	7.0	22.5	52.1
	+	2.0	18.8	17.7
B	–	0.5	9.7	20.3
	+	0.1	0.5	6.1
C	–	13.0	36.3	69.9
	+	0.4	8.2	25.9
D	–	0.2	7.0	40.6
	+	0.6	5.4	28.3

Table 19.6

14 Which variety of apple shows LEAST overall loss due to rotting by fungi when left untreated with fungicide?

15 Which variety of apple shows the greatest reduction in percentage loss due to rotting in July following fungicide treatment?

16 The forecasting of potato blight is based on the occurrence of Smith periods. During a Smith period an air temperature of at least 10°C is accompanied by a relative humidity greater than 90% for at least 11 hours on two consecutive days. The table gives data for a potato field for two weeks.

day	minimum temperature (°C)	number of hours of relative humidity above 90%
1	10	11
2	11	9
3	9	11
4	13	10
5	10	11
6	13	12
7	12	10
8	9	11
9	10	9
10	11	10
11	10	12
12	12	11
13	8	13
14	10	10

Table 19.7

Smith periods occurred on days

A   1–2 and 5–6.   B   1–2 and 11–12.   C   5–6 and 10–11.   D   5–6 and 11–12.

Questions 17 and 18 refer to the following table which shows the mean concentration of a non-biodegradable pesticide residue in the tissues of the organisms in a food chain and the water in their ecosystem.

source of samples	mean concentration of pesticide (ppm)
water	0.00005
plankton	0.04
herbivorous fish	0.23
carnivorous fish	2.07
fish-eating bird	6.00

Table 19.8

17 The mean concentration of pesticide increased by a factor of NINE times between
   A   herbivorous fish and carnivorous fish.
   B   carnivorous fish and fish-eating bird.
   C   plankton and herbivorous fish.
   D   water and plankton.

18 The mean concentration of pesticide in the fish-eating bird is greater than that in the water by a factor of

    A 1.2×10^3 **B** 1.2×10^4 **C** 1.2×10^5 **D** 1.2×10^6

Questions 19 and 20 refer to the following table of results from a survey of breeding success in golden eagles. (Dieldrin is an insecticide that is non-biodegradable).

year	number of nests examined	percentage of nests with broken eggs	percentage of nests suffering theft of eggs	average mass of egg contents per nest (g)	average concentration of dieldrin in egg (ppm)
1	17	5.90	5.90	127	0.03
2	20	10.00	10.00	109	0.19
3	15	13.30	20.00	143	0.53
4	16	18.80	6.23	115	2.11
5	18	27.80	11.10	104	6.63
6	19	31.60	15.80	132	7.15

Table 19.9

19 The data support the hypothesis that an inverse relationship exists between thickness of shell and

    A number of nests examined.

    B average concentration of dieldrin in egg.

    C average mass of egg contents.

    D percentage of nests suffering theft.

20 In which year were five nests found to contain broken eggs?

    A 3   **B** 4   **C** 5   **D** 6

Questions 21 and 22 refer to the following information.

For successful biological control of red spider mites (M) in a glasshouse, the suppliers of *Phytoseiulus* (P), their predator, recommend the use of 200P to deal with a population of 4000M.

21 How many P should be ordered to deal with a population of 28 000M?

    A 70   **B** 700   **C** 1400   **D** 14 000

22 Imagine that a fertilised female P introduced to a glasshouse takes three weeks to lay 50 eggs (25♂ and 25♀) and that each egg takes one week to reach adulthood. How many adult P would be present eight weeks after the introduction of the original female? (Assume that each female lays eggs once and that there are no deaths).

    A 1251   **B** 1276   **C** 1301   **D** 1351

Questions 23 and 24 refer to the following graph. It shows the results of using *Phytoseiulus* as the predator to control red spider mites on cucumber plants in a glasshouse.

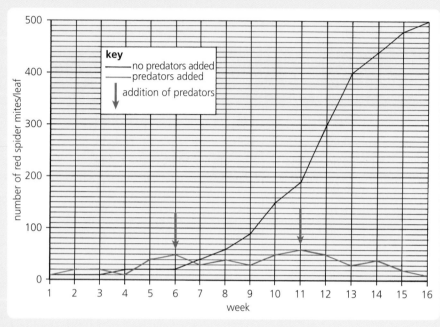

Figure 19.4

23 Starting at week 4, how many more weeks did it take the untreated red spider mites to increase their number per leaf by a factor of 15?

A 4   B 8   C 12   D 16

24 By how many times was the number of red spider mites per leaf lower in the plants treated with the predator compared with the untreated plants at week 14?

A 10   B 11   C 12   D 55

25 Integrated pest management aims to

A eradicate rather than simply control the pests.

B depend on the frequent use of pesticide sprays.

C use chemicals that persist in the environment.

D selectively kill pests without harming useful predators.

20 Animal welfare

Matching test
Match the terms in list X with their descriptions in list Y.

list X
1 altered level of activity
2 ethics
3 ethogram
4 ethology
5 misdirected
6 motivation
7 preference
8 social contact
9 stereotypy
10 wellbeing

list Y
a) behaviour pattern in the form of repetitive movements lacking in variation
b) behaviour that is directed inappropriately towards the animal, another animal or its surroundings
c) type of test that gives an animal a choice between two conditions to determine which one it prefers
d) behaviour involving extreme expression such as hyper-aggression or excessive sleeping
e) animal's quality of life, regarded as acceptable if the animal can behave naturally, grow well and live free of disease
f) moral values and rules that ought to govern human conduct
g) list of all the different observed behaviours shown by an animal
h) study of animal behaviour
i) process that arouses and directs the behaviour of an animal to satisfy one of its basic needs
j) access to company of other members of the animal's own kind

Multiple choice test
Choose the ONE correct answer to each of the following multiple choice questions.

Questions 1 and 2 refer to the following table.

	animal able to:					
	grow free from hunger and thirst	express normal behaviour	live free from fear and distress	resist disease	reproduce and raise offspring	develop free from discomfort, pain or injury
A	✓			✓	✓	
B	✓			✓	✓	✓
C	✓	✓		✓	✓	✓
D	✓	✓	✓	✓	✓	✓

Table 20.1

1 Which line in the table represents the traditional view of wellbeing among domesticated animals?
2 Which line in the table represents the quality of animal welfare based on the freedoms identified by the Farm Animal Welfare Council?
3 Which line in the table below gives the CORRECT answers to blanks 1, 2 and 3 in the following sentence?
   A stereotypy is a ___1___ movement by an animal which expresses ___2___ and lack of ___3___.

	1	2	3
A	hyperactive	motivation	nourishment
B	repetitive	frustration	nourishment
C	hyperactive	motivation	stimulation
D	repetitive	frustration	stimulation

Table 20.2

4 Young dairy calves frolicking in a field in groups is an indication of
   A stereotypy.
   B good welfare.
   C misdirected behaviour.
   D failure of parental control.

5 Many welfare problems that affect commercial farming of domestic fowl are associated with inappropriate pecking. Which of the following practices used to eliminate the problem in an emergency is NOT appropriate for long-term use?
   A use of antipecking sprays
   B setting up of visual barriers
   C periodic use of lighting of low intensity
   D provision of alternative pecking substrates

6 Within a few hours of birth, the members of a litter of piglets compete vigorously with one another during suckling. The sow initiates nursing by emitting grunts. The piglets quickly learn to recognise these sounds as the 'unique voice of their mother'. An investigation compared the gain in mass and level of aggression shown by the piglets of several sows, some the sole occupant of a small pen, others sharing a small pen with other nursing sows. The following graph shows the results.

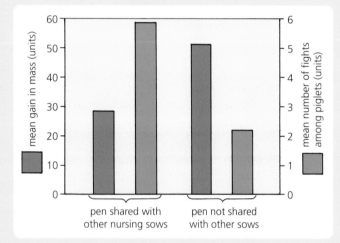

Figure 20.1

A hypothesis that would explain these results is
   A piglets in a noisy pen gain more mass because they benefit from social interaction.
   B piglets in a quiet pen are bored and fight more often with one another.
   C piglets in a noisy pen sometimes fail to hear their mother's grunts and suckle for less time.
   D piglets in a quiet pen fail to gain as much mass because they expend energy to keep warm.

7 The steps given in the following diagram refer to the procedure followed when testing a hypothesis based on an ethogram. Which answer gives the steps in the correct sequence?

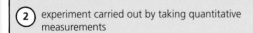

(1) one type of behaviour selected for study

(2) experiment carried out by taking quantitative measurements

(3) ethogram constructed

(4) hypothesis supported or rejected depending on conclusions drawn from results

(5) hypothesis constructed to try to explain chosen behaviour

Figure 20.2

A 1, 3, 5, 2, 4   B 3, 1, 5, 2, 4   C 1, 5, 3, 4, 2   D 3, 1, 5, 4, 2

Questions 8 and 9 refer to the following table. It shows various types of behaviour exhibited by a laboratory mouse in a large cage observed over a period of time divided into 40-second intervals.

time interval	category of behaviour that dominated the time interval			
	moving around cage	feeding	resting	drinking
1	✓			
2	✓			
3		✓		
4		✓		
5		✓		
6				✓
7			✓	
8	✓			
9		✓		
10	✓			
11	✓			
12			✓	

Table 20.3

8 The mouse was most likely to have been motivated to find food during BOTH time intervals

    A 2 and 8.   **B** 3 and 9.   **C** 5 and 8.   **D** 2 and 9.

9 Moving around the cage was the mouse's dominant type of behaviour during a total time of

    A 2 min 30 s.   **B** 3 min 20 s.   **C** 4 min 40 s.   **D** 5 min 10 s.

10 In a preference test, 8 groups of 10 hens were given a choice between joining a large group of other hens in a large space or joining a smaller group of hens in a smaller space. In a second test the choice offered was between joining a larger group of hens in a large space or a smaller group of hens in an equally large space. The tables below show the results.

group	side W	side X
1	6	4
2	5	5
3	6	4
4	4	6
5	7	3
6	5	5
7	6	4
8	4	6

Note:
side W = large space containing 120 hens
side X = small space containing 5 hens

Table 20.4 Test 1

group	side Y	side Z
1	1	9
2	0	10
3	2	8
4	0	10
5	3	7
6	2	8
7	0	10
8	1	9

Note:
side Y = large space containing 120 hens
side Z = large space containing 5 hens

Table 20.5 Test 2

Which of the following conclusions can be correctly drawn from the results?

    A When the size of the space differed, the hens preferred the smaller space.

    B When the number of other hens in the space differed, the hens preferred the smaller space.

    C When the size of the space was kept constant, the hens preferred the smaller group size.

    D When the number of other hens in the space was kept constant, the hens preferred the larger space.

Questions 11 and 12 refer to the following information. Several tests were carried out to investigate the type of food preferred by a large number of pigs. The six tables in the following diagram show the results (S = sorghum, M = maize, Ry = rye and Ri = rice). Each pig was allowed to feed for 300 minutes.

test 1	
mean time spent consuming food (min)	
S	M
118	182

test 2	
mean time spent consuming food (min)	
S	Ry
101	199

test 3	
mean time spent consuming food (min)	
S	Ri
75	225

test 4	
mean time spent consuming food (min)	
M	Ry
120	180

test 5	
mean time spent consuming food (min)	
M	Ri
84	216

test 6	
mean time spent consuming food (min)	
Ry	Ri
102	198

Figure 20.3

11 The foodstuffs arranged in descending order of preference is

A   rice > rye > sorghum > maize.

B   rye > rice > maize > sorghum.

C   rice > rye > maize > sorghum.

D   rye > rice > sorghum > maize.

12 Compared with using a few pigs, use of a large number of pigs makes the results

A   more reliable.

B   fully randomised.

C   free from experimental error.

D   independent of variable factors.

13 Once the primary physiological needs of hunger and thirst have been satisfied, an animal is motivated to seek stimulation as indicated in the following diagram.

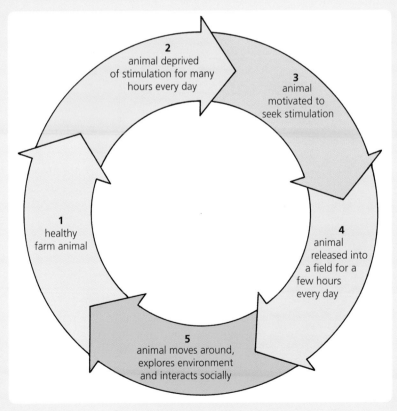

Figure 20.4

If cattle are kept tethered in separate tie-stalls 24 hours per day for several weeks, which arrows in the diagram would no longer apply to the animals' situation?

A 1 and 2　　　**B** 2 and 3　　　**C** 3 and 4　　　**D** 4 and 5

Questions 14 and 15 refer to the following information. An investigation was set up to study the effect of environmental enrichment and size of floor space on harmful antisocial behaviour shown by some young pigs. The following bar chart represents the results.

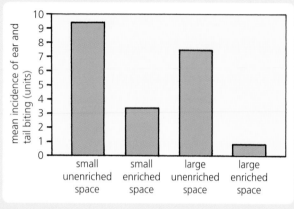

Figure 20.5

14 From this investigation it can be concluded that
   A size of floor space plays a greater role than enrichment of environment in reducing antisocial behaviour.
   B the greatest reduction in antisocial behaviour occurs in small enriched environments.
   C enrichment of environment plays a greater role than size of floor space in reducing antisocial behaviour.
   D the least popular choice by antisocial pigs was the large enriched environment.

15 Four replicates of each treatment were set up in order to
   A exclude uncontrolled variability.
   B minimise the effect of experimental error.
   C give sufficient treatments to enable randomisation.
   D ensure that only one variable factor was being studied.

21 Symbiosis

Matching test

Match the terms in list X with their descriptions in list Y.

list X	list Y
1 chloroplast	a) mode of transmission of parasites which occurs when an infected host physically encounters another host
2 coevolution	b) symbiotic relationship where both partners benefit
3 direct contact	c) symbiotic relationship where one organism benefits at the expense of another organism
4 host	d) ecological relationship between organisms of two different species that live in direct contact with one another
5 interdependence	e) organelle responsible for aerobic respiration thought to have evolved from prokaryotes engulfed by early eukaryotic cells
6 mitochondrion	
7 mutualism	f) photosynthetic organelle thought to have evolved from prokaryotes engulfed by early eukaryotic cells
8 parasite	
9 parasitism	g) organism that carries a parasite from one host to another
10 resistant stage	h) partner that is harmed by loss of resources to a parasite in a symbiotic relationship
11 symbiosis	i) partner that benefits by obtaining resources from a host in a symbiotic relationship
12 vector	j) relationship that exists between the two partners involved in a mutualistic relationship
	k) form of parasite which on being released is able to survive adverse conditions for a long time before meeting a new host
	l) gradual change in parallel of two symbiotic organisms which increases their level of adaptation to suit a dependent or interdependent relationship

Multiple choice test

Choose the ONE correct answer to each of the following multiple choice questions.

1 Symbiosis ALWAYS consists of
   A a mutually beneficial relationship between two members of different species.
   B an intimate relationship between two members of different species.
   C a mutually beneficial relationship between two members of the same species.
   D an intimate relationship between two members of the same species.

2 The following diagram shows a simplified version of the life cycle of *Phytophthora infestans*, the fungus that causes potato blight.

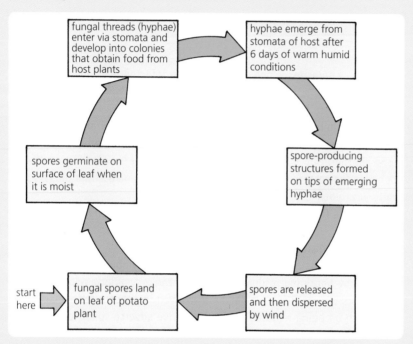

Figure 21.1

Which combination of weather conditions would result in the lowest level of disease transmission?

A warm, windy and humid

B cool, still and humid

C warm, windy and dry

D cool, still and dry

3 The first table below represents the effects of different types of association on the organisms involved. Which line in the second table correctly identifies the three types of association?

association	effect on two organisms involved (+ = gain, − = loss)
1	+/−
2	−/−
3	+/+

Table 21.1

	association		
	1	**2**	**3**
A	parasitism	mutualism	competition
B	mutualism	competition	parasitism
C	parasitism	competition	mutualism
D	competition	parasitism	mutualism

Table 21.2

4 The following table refers to three types of parasite worm.

parasite	mean number of eggs produced per day
hookworm	4×10^4
roundworm	2×10^5
fish tapeworm	1×10^6

Table 21.3

The ratio of mean number of eggs produced per day by fish tapeworm : roundworm : hookworm is

  A 1 : 2 : 4   **B** 1 : 5 : 25   **C** 4 : 2 : 1   **D** 25 : 5 : 1

Questions 5 and 6 refer to the following diagram of the life cycle of a type of liver fluke found in rural parts of Asia.

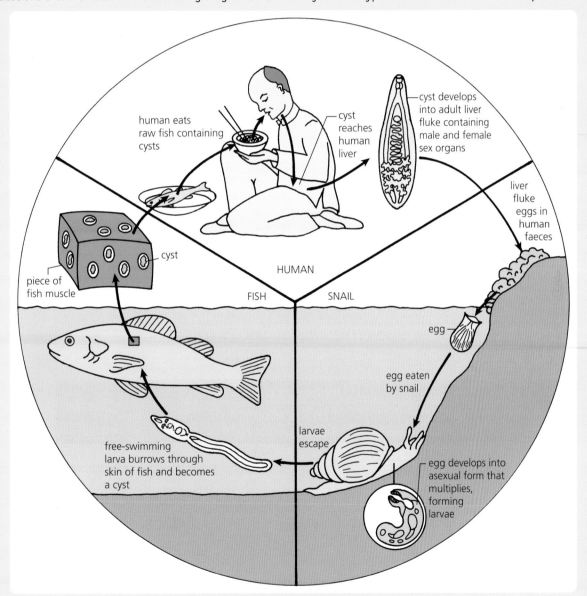

Figure 21.2

5   Which line in the following table is correct?

	primary host(s)	secondary host(s)
A	human	snail and fish
B	snail and fish	human
C	fish and human	snail
D	snail	fish and human

Table 21.4

6   Which of the following procedures would be the easiest and most cost-effective method of breaking this cycle of events?
    A   removing all snails from the fish ponds using molluscicide
    B   restocking the ponds with fish that eat snails
    C   cooking all fish thoroughly before consumption by humans
    D   building sewage treatment works beside the fish ponds

7   It is correct to say that a parasite ALWAYS
    A   needs continuous contact with its host to survive.
    B   gives the host a disease.
    C   benefits at the expense of the host.
    D   eventually kills its host.

8   The first table below refers to the methods by which five parasites are spread.
    Which line in the second table is correct?

parasite		method of spread
1	yellow fever virus	carried by female mosquitoes from host to host
2	scabies mite	infected skin or contaminated clothes touch the skin of new host
3	tapeworm	cysts present in undercooked meat are consumed
4	AIDS virus	potential host engages in unprotected sex with infected host
5	rabies virus	carried in saliva of rabid dog from host to host

Table 21.5

	mechanism of transmission		
	vector	release of resistant stage	direct contact
A	1 and 5	3	2 and 4
B	1	3 and 4	2 and 5
C	2 and 3	1	4 and 5
D	4	2 and 5	1 and 3

Table 21.6

Questions 9 and 10 refer to the following information. An experiment was set up to investigate the effect of a digestive enzyme from the primary host on the hatching of tapeworm cysts. The results are shown in the following graph.

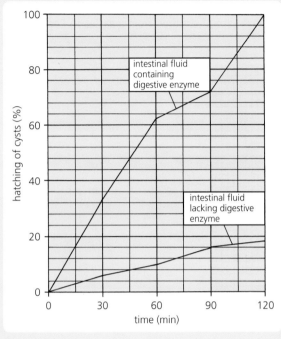

Figure 21.3

9 From the graph it can be concluded that at 1 h 30 min, the presence of the enzyme compared to the lack of the enzyme brought about the hatching of more cysts by a factor of

   A 3.5 times.   B 4.5 times.   C 5.5 times.   D 56.0 times.

10 In this experiment the dependent variable factor was

   A presence of digestive enzyme.
   B absence of digestive enzyme.
   C time in minutes.
   D percentage hatching of cysts.

11 Which of the following is LEAST likely to be an adaptation possessed by a parasite that lives in the gut of its primary host?

A suckers   **B** large surface area   **C** long intestine   **D** rows of hooks

Questions 12 and 13 refer to the following graph which shows the death rate of children from measles in England and Wales many years ago.

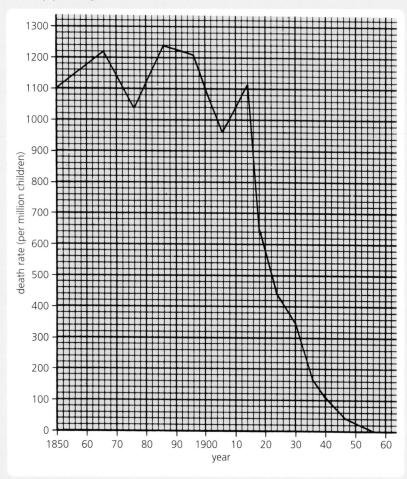

Figure 21.4

12 By how many times was the death rate greater in 1890 than in 1930?

A 2.5   **B** 3.0   **C** 3.5   **D** 4.0

13 Compared with the death rate in 1900, what percentage reduction had occurred by 1940?

A 10   **B** 90   **C** 99   **D** 110

14 In 1890 the German scientist, Robert Koch, set out his famous criteria for deciding whether or not a given bacterium is the cause of a given disease. These so-called postulates (necessary conditions) are shown in the diagram below.

① The scientist must be able to recover the bacteria from the experimentally infected host.

② The scientist must be able to isolate the bacteria from any of these hosts and grow the bacteria in pure culture.

③ The bacteria must be present in every host suffering the disease.

④ It must be possible to reproduce the disease by inoculating a pure culture of the bacterium into a healthy susceptible host.

Figure 21.5

Their correct order is

A 2, 3, 1, 4   B 2, 3, 4, 1   C 3, 2, 1, 4   D 3, 2, 4, 1

Questions 15 and 16 refer to the information in the following passage.

Aphids have numerous enemies including a type of wasp which lays its eggs inside the aphid's body. When the eggs hatch, the grubs feed on the aphid. However, many aphids survive and may act as vectors for numerous viruses that cause diseases of green plants. The viruses gain access to the plant while an aphid is using its piercing mouthparts to extract sugary juice from the plant's phloem tissue as shown in the diagram.

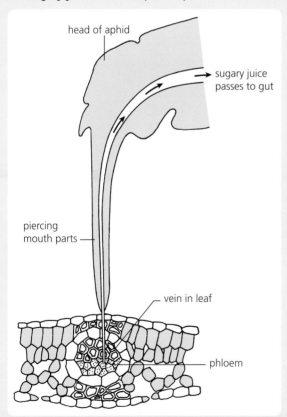

head of aphid

sugary juice passes to gut

piercing mouth parts

vein in leaf

phloem

Figure 21.6

Some types of aphid excrete sweet 'honeydew' which is used as food by a type of ant. These insects gather aphid eggs in autumn and tend them during the winter. When spring arrives, the ants transport the newly hatched aphids to suitable plants where they can begin feeding. The ants are then able to 'milk' the aphids for their 'honeydew'.

15 Which line in the following table is an example of parasitism?

	host	parasite
A	green plant	virus
B	aphid	green plant
C	wasp	aphid
D	aphid	virus

Table 21.7

16 The relationship shown by the aphid and the ant is an example of

A mutualism.   B parasitism.   C predation.   D competition.

17 The Nile crocodile allows the Egyptian plover bird to pick out leeches attached to its gums as shown in the diagram.

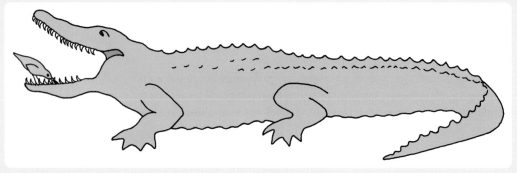

Figure 21.7

In this relationship
A the crocodile is the host and the bird is the parasite.
B the bird is the host and the leech is the parasite.
C the crocodile is the host and the leech is the parasite.
D only one organism, the crocodile, receives any benefit.

18 A mycorrhiza is an intimate mutualistic association between the roots of a plant and one or more fungi. The table below shows the results of an experiment involving seedlings of several types of tree grown for three months.

tree	mean dry mass of seedlings (mg)	
	non-mycorrhizal	mycorrhizal
beech	868	1 163
eucalyptus	7 812	10 624
oak	1 151	1 565
pine	318	438

Table 21.8

The type of tree that showed the greatest percentage increase in mean dry mass when in an association with the mycorrhizal fungus was
A beech.   B eucalyptus.   C oak.   D pine.

19 Every member of a type of hermit crab called *Pagurus* (see diagram) has a sea anemone (*Adamsia*) on its shell with the latter's mouth and stinging tentacles close to the mouth of the crab. If the crab is removed from its shell, the anemone drops off and dies. If the anemone is forcibly removed from the shell, the crab soon replaces it with another using its pincers.

Figure 21.8

In this relationship
A the crab obtains scraps of food while the anemone gains protection.
B the anemone obtains scraps of food while the crab gains protection.
C the crab obtains nitrogenous compounds while the anemone gains extra oxygen.
D the anemone obtains nitrogenous compounds while the crab gains extra oxygen.

Questions 20, 21 and 22 refer to the following table.

relationship	benefits gained by:	
	partner 1	partner 2
A	supply of ammonium compounds needed to make protein	supply of sugar and use of a stable habitat
B	supply of sugar and oxygen	supply of minerals, water and a means of transport
C	digestion of plant cell walls present in its food	supply of sugar and provision of a habitat
D	transfer of pollen grains to stigmas	supply of seeds to feed larvae

Table 21.9

20 Which relationship's partners are yukka (a flowering plant) and the yukka moth?
21 Which relationship involves clover (a leguminous plant) and nitrogen-fixing bacteria?
22 Which relationship involves the fungus and the alga that make up a lichen?

23 Chloroplasts are thought have evolved by endosymbiosis. The boxes in the diagram below give some of the stages thought to have occurred in this process.

① over a very long period of time the two endosymbionts become a single organism with its parts inseparable

② eukaryotic cell fails to digest the cyanobacterium which multiplies and gradually evolves into chloroplasts

③ eukaryotic cell engulfs a photosynthetic cyanobacterium

④ a form of mutualism develops where the host cell provides carbon dioxide and the chloroplasts provide sugar and oxygen

Figure 21.9

Their correct sequence is

A 2, 3, 1, 4   B 2, 3, 4, 1   C 3, 2, 1, 4   D 3, 2, 4, 1

24 The following statements present evidence for the hypothesis of the endosymbiotic origin of mitochondria and chloroplasts. Which one contains information that is NOT correct?

Chloroplasts and mitochondria

A have an inner membrane bearing enzymes and transport systems similar to those in bacterial membranes.

B contain ribosomes more similar to those found in eukaryotes than those present in prokaryotes.

C have their own DNA which is circular like that found in prokaryotes.

D replicate by a method that is reminiscent of the process of cell division in bacteria.

25 The table below refers to growth factors required by two species of micro-organism if they are to be successfully cultured.

species	factor required by micro-organism for growth	metabolite synthesised and released by growing micro-organism
Mucor ramannianus	thiazole	pyrimidine
Rhodotorula rubra	pyrimidine	thiazole

Table 21.10

Which of the following relationships would be most likely to occur if these two micro-organisms were cultured together?

A mutualism   B competition   C incompatibility   D parasitism

22 Social behaviour

Matching test

Match the terms in list X with their descriptions in list Y.

list X

1 appeasement
2 alliance
3 altruism
4 co-operative hunting
5 dominant
6 ecosystem service
7 grooming
8 keystone
9 kin selection
10 parental care
11 social group
12 social hierarchy
13 social status
14 taxonomic
15 threat display
16 worker

list Y

a) an individual's level of rank in a social hierarchy
b) unit consisting of several members of a species who live together and respond to one another
c) graded order of rank among members of a social group resulting from aggressive behaviour
d) benefit gained by humans from processes carried out by members of an ecosystem's community
e) type of species that plays a critical role in the structure and working of an ecosystem
f) type of social behaviour used by a member of a social group to concede defeat to a dominant rival
g) ritualised behaviour used by a member of a social group to intimidate a rival without engaging in a real fight
h) member of a social group able to intimidate all other members without being attacked in return
i) term referring to a group of living things classified together because they have certain characteristics in common
j) close relationship among female primates in a social group where reproductive success can increase social status
k) process by which primates clean one another's fur while reducing tension within their social group
l) type of social behaviour where adult primates provide their offspring with food, protection and transport
m) non-reproductive member of an insect colony that co-operates with close relatives to ensure the survival of the young
n) process which favours acts of apparent altruism carried out to help close relatives
o) unselfish behaviour which harms the donor but benefits the recipient
p) method of obtaining food that benefits subordinate animals in addition to the group's dominant leader

→

Multiple choice test

Choose the ONE correct answer to each of the following multiple choice questions.

1 The diagram shows three different postures that may be adopted by a male greylag goose (gander) in the presence of another greylag gander.

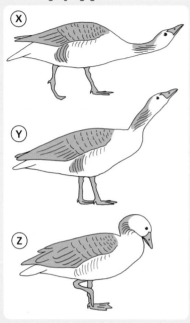

Figure 22.1

Which line in the table correctly identifies the message conveyed in each case?

	threat being indicated at a distance	subordinate social approach being offered	all-out attack about to be launched
A	Y	X	Z
B	X	Z	Y
C	Y	Z	X
D	Z	Y	X

Table 22.1

2 The boxed list in the following diagram refers to pecking behaviour observed among six hens (P, Q, R, S, T and U).

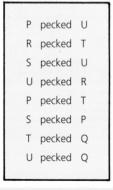

P pecked U
R pecked T
S pecked U
U pecked R
P pecked T
S pecked P
T pecked Q
U pecked Q

Figure 22.2

Which bird was THIRD in the peck order?

A P   B U   C R   D T

3 Which of the following does NOT occur as a direct result of a herd of animals being dominated by one older male?
  A minimum aggression among the group members
  B experienced leadership during times of crisis
  C promotion of the group's chance of survival
  D equal choice of food for all herd members

4 Which of the following is a social mechanism for defence?
  A alarm calls in birds
  C poison glands in snakes
  B spines on bodies of hedgehogs
  D foul-smelling secretions squirted by skunks

Questions 5 and 6 refer to the following information. Vampire bats need regular feeds of blood. In the absence of food, they reach starvation point at about 60 hours from their previous meal as shown in the accompanying graph. Bat V has just fed on the blood of a host.

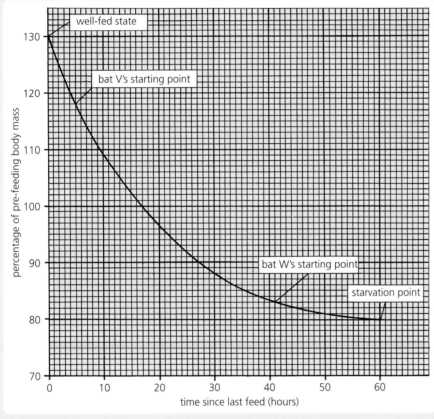

Figure 22.3

5 Bat V now regurgitates a donation of blood to bat W which results in bat W's body mass increasing by 7% and bat V's body mass decreasing by 13%.
  Which line in the following table is correct?

	time gained by bat W (h)	time lost by bat V (h)
A	7	13
B	8	14
C	13	7
D	14	8

Table 22.2

6 Two days later the roles are reversed and bat W donates blood to bat V. This form of social behaviour is called

  A kin selection. **B** reciprocal altruism. **C** social hierarchy. **D** ecosystem servicing.

Questions 7 and 8 refer to the following two tables. The first shows a system of scoring points when playing the prisoner's dilemma game. The second shows the results of students P and Q playing 20 rounds of the game.

	P remains silent	**P betrays Q**
Q remains silent	P and Q each receive 3 points	P receives 5 points, Q receives 0 points
Q betrays P	Q receives 5 points, P receives 0 points	P and Q each receive 1 point

Table 22.3

round	student P's score	student Q's score
1	1	1
2	5	0
3	1	1
4	3	3
5	0	5
6	5	0
7	1	1
8	0	5
9	3	3
10	1	1
11	1	1
12	3	3
13	3	3
14	5	0
15	0	5
16	3	3
17	5	0
18	0	5
19	1	1
20	3	3

Table 22.4

7 During which rounds did Q betray P, who remained silent?

  A 1, 2, 3, 7, 10, 11, 19 **B** 2, 6, 14, 17 **C** 4, 9, 12, 13, 16, 20 **D** 5, 8, 15, 18

8 Which of the following rounds could ALL be interpreted as P punishing Q for a betrayal?

  A 2, 6, 9, 19 **B** 3, 7, 11, 19 **C** 2, 6, 11, 19 **D** 3, 6, 11, 20

9 The following diagram represents the process of kin selection.

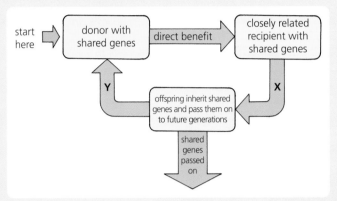

Figure 22.4

Which line in the table below correctly identifies X and Y?

	X	Y
A	shared genes passed on	direct benefit
B	indirect benefit	shared genes passed on
C	shared genes passed on	indirect benefit
D	indirect benefit	shared genes passed on

Table 22.5

10 The following diagram shows the four possible steps in the evolution of insect colonies.

① Some insects in the group are reproductive and other non-reproductive individuals carry out tasks that benefit the whole community.

② The insects as a community co-operate in the building of a nest but each female raises her own brood of young.

③ Only a few individuals in the community are reproductive and the vast majority of non-reproductive individuals are specialised to perform tasks that ensure survival of young.

④ Community of insects build a nest communally and all co-operate in the rearing of the broods of young.

Figure 22.5

Their correct order is

A 2, 4, 1, 3   B 4, 2, 1, 3   C 2, 4, 3, 1   D 4, 2, 3, 1

11 The stick graph below shows the percentage time spent by a worker bee patrolling the hive during the first three weeks of her life.

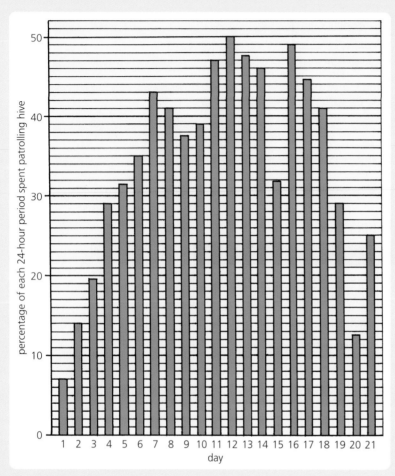

Figure 22.6

How many more hours did she spend patrolling the hive on day 12 compared with day 20?

A 4   B 9   C 12.5   D 37.5

12 Which of the following is an activity carried out only by drone bees?

A fertilising the queen   B feeding the larvae

C guarding the hive   D foraging for food

13 Which of the following is NOT an example of an essential ecosystem service provided by a keystone species?

A decomposition of plant debris by termites

B pollination of flowers by bumble bees

C control of pests by parasitic wasps

D laying of eggs on cabbages by butterflies

14 The table below shows the result of an investigation into some aspects of parenthood in two types of primate.

primate	mean age of mother at first birth (years)	mean weaning age (years)	mean interbirth interval (years)	mean age of youngster at independence (years)
chimpanzee	14.3	5.3	5.9	5.3
human forest dweller	19.7	2.5	3.1	18.0

Table 22.6

Which of the following conclusions can be correctly drawn from the data?

A The interval of time from the birth of one infant to that of the next is longer among humans.

B Suckling of youngsters finishes earlier among human forest dwellers than chimpanzees.

C The length of time that elapses between the birth of infants is directly related to the mother's age.

D Chimpanzees wean their youngsters on to solid food earlier than human forest dwellers.

15 Which of the following statements is NOT correct?

In primates, parental care allows the learning, during play, of complex social behaviour such as

A strong handgrip.

B sharing of resources.

C co-operation with peers.

D communication skills.

16 Which of the following are BOTH examples of social behaviour in primates that contribute to the reduction of unnecessary conflict between the members of a social group?

A appeasement and rejection of sexual advances

B ritualistic threat display and appeasement

C grinning facial expression and rejection of grooming

D servile body posture and lack of appeasement

17 The scatter graph below shows the results of a survey of the percentage of daytime hours devoted to social grooming by groups of Old World primates.

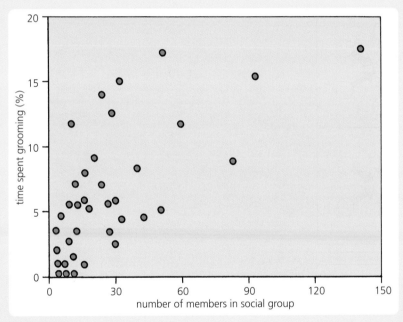

Figure 22.7

Which of the following conclusions can be correctly drawn from these results?

A As the size of a social group increases, the time spent by its members on grooming decreases.

B The members of small-sized social groups spend more time grooming than those in larger groups.

C The larger the size of a social group, the greater the percentage time spent on grooming.

D The more densely populated a social group, the more time its members spend on grooming.

18 The following diagram represents four female vervet monkeys (W, X, Y and Z) and the grooming alliances formed with other females before and after childbirth.

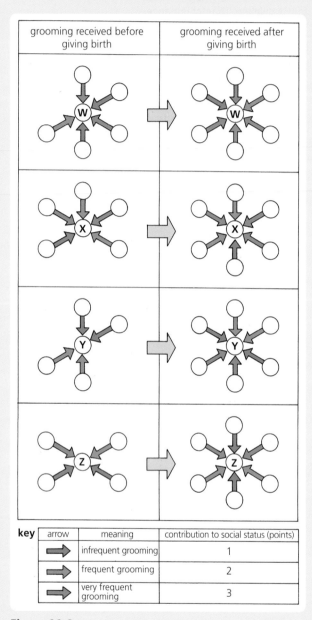

key	arrow	meaning	contribution to social status (points)
	⇒	infrequent grooming	1
	⇒	frequent grooming	2
	⇒	very frequent grooming	3

Figure 22.8

Which vervet monkey received the greatest INCREASE in social status following childbirth?

A W   B X   C Y   D Z

Questions 19 and 20 refer to the following table which compares the percentage time spent on various activities by three types of primate.

activity	time spent on activity (%)		
	howler monkey	spider monkey	capuchin
travelling in search of food	6.72	18.74	25.31
feeding on fruit	3.71	20.41	23.47
feeding on leaves	6.37	2.96	0.38
feeding on flowers	2.93	2.64	0.06
feeding on insects	0.00	0.35	4.88
total for all types of food	13.01	26.36	28.79

Table 22.7

19 Which of the following conclusions can be correctly drawn from the data?
  A Capuchins spend less time travelling between food sites than spider monkeys.
  B Spider monkeys spend more time eating insects than capuchins.
  C Howler monkeys spend less time eating leaves than spider monkeys.
  D Spider monkeys spend more time feeding on flowers than capuchins.

20 When comparing howler and spider monkeys, which of the following modes of feeding shows an increase in percentage time spent on it by a factor of 5.5?
  A feeding on fruit
  B feeding on leaves
  C feeding on flowers
  D feeding on insects

23 Mass extinction and biodiversity

Matching test

Match the terms in list X with their descriptions in list Y.

list X
1 biodiversity
2 degradation
3 dominant
4 ecosystem diversity
5 extinction
6 extinction rate
7 genetic diversity
8 habitat destruction
9 habitat island
10 mass extinction event
11 megafauna
12 species diversity

list Y
a) number of distinct ecosystems present in a defined area
b) state found among the members of a population resulting from the genetic variation present in their DNA
c) feature of an ecosystem based on the richness of its species and their relative abundance
d) area surrounded by a dissimilar ecosystem that cannot be colonised by the enclosed area's species
e) human activity leading to possible extinction of many species
f) reduction in the quality of a natural ecosystem as a result of human activity
g) the largest terrestrial animals belonging to a region or a period of time
h) most prevalent species that determines the appearance and composition of the community
i) total variation that exists among all living things on Earth
j) irreversible loss of a species of living things from planet Earth
k) measure of the number or percentage of species irreversibly lost per unit time
l) disruptive occurrence that changes the global environment and causes species to perish

Multiple choice test

Choose the ONE correct answer to each of the following multiple choice questions.

1 About 440 million years ago the world was dominated by warm seas and much of the land mass was submerged. Five million years later, large ice sheets had formed, sea levels had dropped and vast areas of land had become exposed. Which of the following types of organism would be MOST likely to suffer a wave of mass extinction under such circumstances?
   A plankton living at the surface of the ocean
   B deep sea invertebrate species
   C marine animals adapted to warm shallows
   D seaweeds native to cold waters

2 Which line in the table below indicates the series of events triggered by the lengthy volcanic eruptions that are thought to have caused the Permian mass extinction event of 250 million years ago?

	volume of CO_2 in the atmosphere	temperature of the seas	quantity of oxygen dissolved in the seas
A	increased	increased	decreased
B	decreased	decreased	increased
C	increased	decreased	decreased
D	decreased	increased	increased

Table 23.1

3 At the beginning of the 21st century, all of the following animals were in danger of becoming extinct EXCEPT
   A red deer.　　B blue whale.　　C Siberian tiger.　　D mountain gorilla.

4 The process of extinction of species is decelerated by
   A sudden climatic changes.
   B overhunting of humans.
   C habitat destruction.
   D environmental conservation.

5 The diagram below represents four populations of a species where each symbol represents a diploid individual possessing two alleles of one gene.

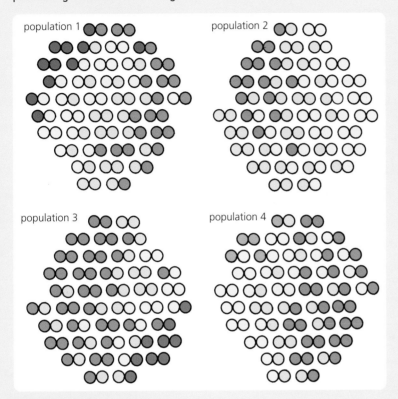

Figure 23.1

Which of the following statements is correct?

A   Population 4 is the most genetically diverse of the four populations.

B   Three of the alleles are each found to be present in only one population.

C   The purple allele is the rarest and only appears in population 1.

D   The red allele occurs most frequently in population 3 and least frequently in population 1.

6 Species diversity comprises BOTH the

A   richness of species and the proportion of each species in an ecosystem.

B   proportion of each species and the health of the plant community in an ecosystem.

C   health of the plant community and the dominant species in an ecosystem.

D   dominant species and the richness of species in an ecosystem.

7 The table below compares four communities from four different ecosystems. Which has the greatest species diversity?

species	relative abundance (%)			
	community A	community B	community C	community D
P	20	10	15	30
Q	10	60	20	20
R	0	10	15	15
S	15	5	15	20
T	50	10	20	0
U	5	5	15	15

Table 23.2

8 Ecosystem diversity refers to the number of distinct
   A species present in a defined ecosystem.
   B species present in a group of related ecosystems.
   C ecosystems present in a defined area.
   D ecosystems present in a random selection of locations.

9 Which of the following is NOT a habitat island?
   A cold mountaintop surrounded by warm lowland
   B pocket of natural forest surrounded by farmland
   C freshwater lake surrounded by dry land
   D clearing in the jungle created by fallen trees

10 The following diagram shows four new uninhabited islands that have been formed simultaneously. After a few years, which island is most likely to possess the lowest level of species diversity?

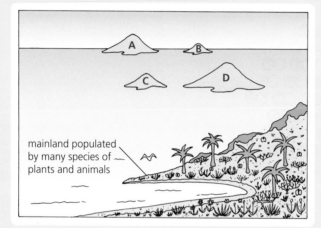

mainland populated by many species of plants and animals

Figure 23.2

24 Threats to biodiversity

Matching test
Match the terms in list X with their descriptions in list Y.

list X
1 bottleneck effect
2 bottleneck event
3 climate change modelling
4 conservation
5 greenhouse effect
6 habitat corridor
7 habitat fragmentation
8 introduced
9 invasive
10 naturalised
11 overexploitation
12 recolonization

list Y
a) naturalised species that has spread rapidly and has outcompeted native species
b) foreign species that has been cultivated intentionally or released accidentally by human activities
c) non-native species that has established itself within wild communities
d) removal and use of species at a rate that exceeds their maximum rate of reproduction
e) return to and resettlement in a habitat after local extinctions
f) cause of the bottleneck effect, typically a natural disaster
g) protection and careful management of natural resources
h) global warming by infrared rays trapped and returned to Earth by certain gases
i) computer programmes that use quantitative methods to simulate the effect of various factors on climate
j) narrow strip of quality habitat by which a species can move between disconnected habitat fragments
k) formation of several fragments whose total surface area is less than that of the original habitat
l) result of the wiping out of a significant percentage of a population and its genetic diversity, leaving it ill-equipped to adapt to environmental change

Multiple choice test
Choose the ONE correct answer to each of the following multiple choice questions.

1 The following graph shows the results of a survey on the quantity of wild sea bass of breeding age present in the seas around England.

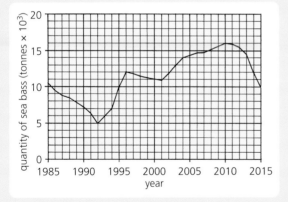

Figure 24.1

Which row in the following table is correct?

	percentage increase between 1992 and 1996	percentage decrease between 2010 and 2014
A	120	25.0
B	120	37.5
C	140	25.0
D	140	37.5

Table 24.1

2 The accompanying graph charts the effect of increased intensity of fishing on four species of edible fish caught in the North Sea. (Sustained yield means the maximum yield of fish that can be maintained from year to year.) Which species is least affected by increased intensity of fishing?

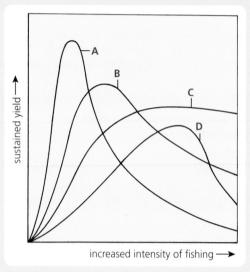

Figure 24.2

3 Which of the following is NOT a measure adopted by European countries to promote the recovery of depleted marine fish stocks in the North Sea?
  A The number of fishing boats is controlled.
  B The length of time allowed at sea is now reduced.
  C The areas where fishing is allowed are limited.
  D The fishing is restricted to shallow marine waters.

Questions 4 and 5 refer to the following information. The effective size (N_e) of a population can be calculated using the formula:

$$N_e = \frac{4N_fN_m}{N_f + N_m}$$

where N_f = number of females that successfully breed
and N_m = number of males that successfully breed.
The following table gives data for four populations W, X, Y and Z, each of which possesses a total of 2000 individuals.

population	N_f	N_m
W	900	600
X	800	700
Y	650	850
Z	950	550

Table 24.2

4 Which population has the highest effective size?
  A W   B X   C Y   D Z

5 What is the effective size of population W expressed as a percentage of the total population?
  A 56   B 60   C 72   D 90

Questions 6, 7 and 8 refer to the following information. The table lists several species of catfish found in South America.

scientific name	common name	habitat	colour code in pie chart diagram
Pseudoplatystoma corruscans	spotted sorubim catfish	fresh water	
Pseudoplatystoma tigrinum	tiger sorubim catfish	fresh water	
Pseudoplatystoma fasciatum	barred sorubim catfish	fresh water	
Genidens barbus	white sea catfish	salt water	
Netuma thalassina	giant marine catfish	salt water	

Table 24.3

The pie charts in the following diagram represent the results of analysing samples from packets of fresh fish products on sale in markets to establish the identity of the fish species they contain.

product 1   product 2   product 3   product 4

Figure 24.3

6   The following list gives the steps involved in the identification procedure.
    1 amplified DNA cut up using a restriction enzyme
    2 result compared with standard species-specific DNA profiles
    3 DNA fragments separated by gel electrophoresis to give 'bar code'
    4 DNA from sample of unknown fish amplified using PCR
    The correct order in which these steps would be carried out is
    A 1, 4, 2, 3   **B** 1, 4, 3, 2   **C** 4, 1, 2, 3   **D** 4, 1, 3, 2

7   If all four of the fish products were labelled 'Catfish (*Pseudoplatystoma corruscans*)' which product(s) would be mislabelled?
    A 1, 2, 3 and 4   **B** 2, 3 and 4   **C** 3 and 4   **D** 4 only

8   If all four of the fish products were labelled 'Catfish (sorubim)' which product(s) would be mislabelled?
    A 1, 2, 3 and 4   **B** 2, 3 and 4   **C** 3 and 4   **D** 4 only

9   Which of the following graphs BEST illustrates the aftermath of a bottleneck event on a population?

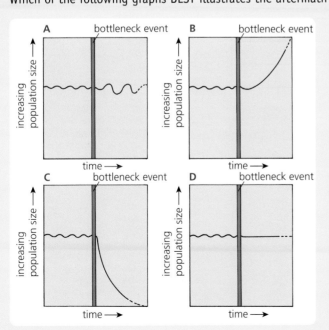

Figure 24.4

10 Which line in the table correctly refers to a population that has suffered the bottleneck effect?

	quantity of genetic diversity now present	allele frequency compared with the original population	number of genes present in genome
A	increased	unchanged	unchanged
B	decreased	different	unchanged
C	increased	unchanged	reduced
D	decreased	different	reduced

Table 24.4

11 The following table compares a habitat fragment with a large area of habitat. Which line is correct?

	habitat fragment	large area of habitat
A	less susceptible to degradation at its edges	more susceptible to degradation at its edges
B	supports lower species richness	supports higher species richness
C	lower chance of edge species invading habitat at expense of interior species	higher chance of edge species invading habitat at expense of interior species
D	lower ratio of total length of edge to total surface area of interior	higher ratio of total length of edge to total surface area of interior

Table 24.5

12 The following list gives a description of three types of non-native species.
   X = species which has spread rapidly and eliminated native species.
   Y = species which has been brought into the country accidently by humans and allowed to grow.
   Z = species which has established itself within wild communities without detrimental effects.
   Which line in the following table correctly matches the types of species with their descriptions?

	type of species		
	naturalised	invasive	introduced
A	Z	X	Y
B	X	Z	Y
C	Z	Y	X
D	Y	X	Z

Table 24.6

13 Which of the following statements is NOT correct?
   Invasive species are often very successful because they
   A outcompete indigenous populations for resources.
   B are the natural prey of native predatory species.
   C hybridise with a native species in its ecosystem.
   D are living free of their natural predators and pathogens.

14 The following diagram shows a simplified version of the greenhouse effect as it affects the Earth at present.

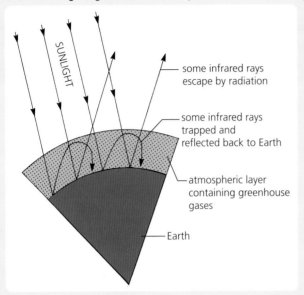

Figure 24.5

Which of the following diagrams BEST represents the situation envisaged for the future if current outputs of greenhouse gases continue?

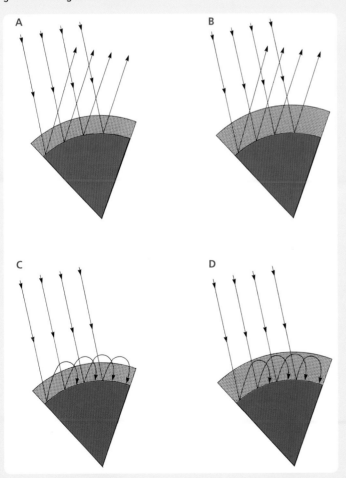

Figure 24.6

15 Which double row in the following table correctly compares the typical responses of two types of species to climate change?

	species type	More able to adapt quickly to climate change?	More highly specialised for life in a specific ecosystem?	More tolerant of a wide range of climatic conditions?	More likely to face extinction?
A	generalist	yes	no	yes	no
	specialist	no	yes	no	yes
B	generalist	no	yes	no	yes
	specialist	yes	no	yes	no
C	generalist	yes	no	no	yes
	specialist	no	yes	yes	no
D	generalist	no	yes	yes	no
	specialist	yes	no	no	yes

Table 24.6

Specimen Examination 1

Choose the ONE correct answer to each of the following multiple choice questions.

1 The following diagram shows a molecule of DNA before replication has occurred.

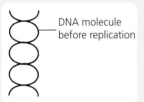
DNA molecule before replication

Figure 1

If ——————— represents an original DNA strand and - - - - - - represents a new DNA strand, which of the following daughter DNA molecules will result from replication of the DNA molecule shown above?

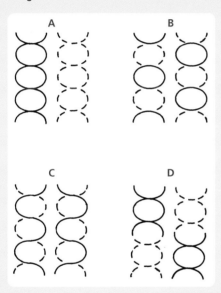

Figure 2

2 The set of results below shows an analysis of the DNA bases contained in the cells of a cow's thymus gland.

base composition (%)			
X	guanine	Y	Z
28.7	21.5	21.3	28.4

Table 1

Which of the following is a possible correct identification of the bases?

	X	Y	Z
A	cytosine	adenine	thymine
B	thymine	adenine	cytosine
C	adenine	cytosine	thymine
D	cytosine	thymine	adenine

Table 2

3   An mRNA template is
    A   translated from protein.   B   transcribed into protein.
    C   translated into DNA.   D   transcribed from DNA.

4   The following list gives the steps that may be used in the future to culture and make use of human stem cells.
    1 stem cells cloned into colonies in the laboratory
    2 differentiated cells used to repair damaged organs
    3 stem cells induced by chemical means to differentiate
    4 undifferentiated cells extracted from embryo
    Which of the following is the correct sequence of steps?
    A   4, 1, 3, 2   B   4, 1, 2, 3   C   1, 4, 3, 2   D   1, 4, 2, 3

5   What name is given to the type of point mutation where one incorrect nucleotide occurs in place of the correct nucleotide in a DNA chain?
    A   deletion   B   insertion   C   inversion   D   substitution

6   Based on genetic evidence, experts support the theory that living things are made up of three main domains called
    A   bacteria, fungi and eukaryotes.
    B   prokaryotes, plants and eukaryotes.
    C   bacteria, archaea and eukaryotes.
    D   bacteria, archaea and vertebrates.

7   The following diagram shows ways in which molecules may move into and out of a respiring animal cell. Which of these could be active uptake of potassium ions?

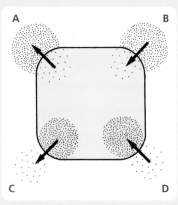

Figure 3

8   The following diagram shows a metabolic pathway controlled by end product inhibition.

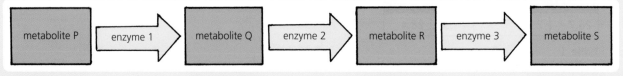

Figure 4

    Metabolite S would bring about this process of end product inhibition by interacting with
    A   metabolite P.   B   enzyme 1.   C   metabolite R.   D   enzyme 3.

9   Which of the following is the correct sequence of the processes that occur during cellular respiration?
    A   glycolysis → citric acid cycle → electron transport
    B   citric acid cycle → glycolysis → electron transport
    C   glycolysis → electron transport → citric acid cycle
    D   electron transport → citric acid cycle → glycolysis

10 Maximum oxygen uptake (VO_2 max) is regarded as the best indicator of a person's
   A fitness and it improves with training.
   B fitness and it increases with age.
   C strength and it improves with training.
   D strength and it increases with age.

11 An organism that is able to maintain a steady state in its internal environment despite changes in its external environment is called
   A an effector.   B a regulator.   C a conformer.   D a controller.

12 Following overheating of a mammal's internal environment, the skin acts as an effector. Which line in the following table correctly indicates the changes that occur?

	altered state of arterioles leading to skin	altered state of erector muscles in skin
A	dilated	relaxed
B	constricted	relaxed
C	dilated	contracted
D	constricted	contracted

Table 3

13 The diagram below shows four imaginary animals. Which one is best suited to an extremely cold climate?

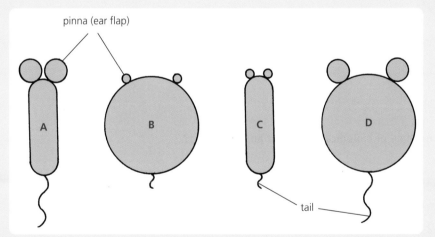

Figure 5

14 The following diagram shows a piece of human DNA ready to be sealed into a bacterial plasmid.

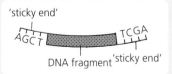

Figure 6

Which of the diagrams below represents a suitable plasmid?

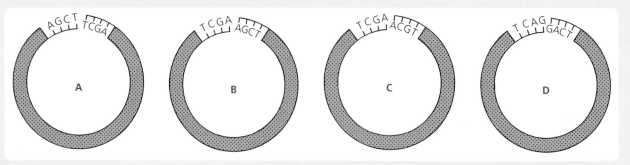

Figure 7

15 Which area in the following diagram represents a region of the world where food of sufficient quality and quantity is available but where most of the local people cannot afford to buy it?

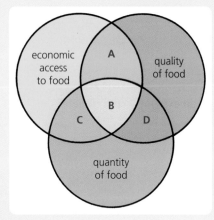

Figure 8

16 Which of the following is NOT an example of biological control of an insect pest?
   Introduction of
   A a parasite to the insect pest.
   B one of the pest's natural predators.
   C a leguminous plant among the crop under attack.
   D a bacterium that is pathogenic to the pest.

17 In order to defend the hive, a worker bee will sting an intruder even although this will result in the death of the worker bee. This form of behaviour is called
   A altruism.   B appeasement   C threat display.   D social hierarchy.

18 Which of the following is NOT a social mechanism for defence?
   A mobbing employed by musk oxen when under attack
   B dominance hierarchy observed by baboons when on the move
   C injection of poison into enemies by adders when threatened
   D circular formation adopted by bobwhite quails when resting

19 Which row in the table below is true of almost all groups of primates?

	ecological niche	benefit to group
A	herbivorous	continuous availability of food throughout the year
B	omnivorous	exploitation of a wide variety of foods
C	carnivorous	co-operative hunting and sharing of kill
D	insectivorous	ease of capture of ants and termites using tools

Table 4

20 Habitat fragmentation involves phenomena such as:

1 reduction in the total area of habitat

2 increase in ratio of total length of edges to total surface area of interior

3 decrease in average size of each fragment with time

Which of these are illustrated in the accompanying diagram?

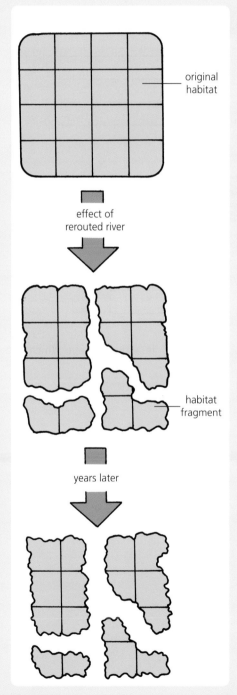

Figure 9

A 1 only   **B** 1 and 2   **C** 2 and 3   **D** 1, 2 and 3

Choose the ONE correct answer to each of the following multiple choice questions.

1   The following table compares the characteristics of two types of cell.
    Which row in the table is NOT correct?

	characteristic	animal cell (eukaryote)	bacterial cell (prokaryote)
A	true nucleus with double membrane	present	absent
B	arrangement of DNA	linear	circular
C	plasmids	present	absent
D	ribosomes	present	present

Table 1

2   Which line in the following table represents DNA?

	relative percentage of base				
	adenine	guanine	uracil	cytosine	thymine
A	23	27	0	27	23
B	23	27	0	23	27
C	29	21	24	26	0
D	29	29	0	21	21

Table 2

3   The flow chart in the accompanying diagram refers to the coding for and synthesis of an active protein. Which arrow represents the process of translation?

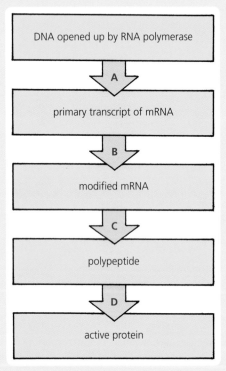

Figure 1

4   Which row in the table correctly describes meristematic cells?

	location of cells	state of cells
A	animals	differentiated and specialised
B	plants	undifferentiated and unspecialised
C	animals	undifferentiated and unspecialised
D	plants	differentiated and specialised

Table 3

5   The following diagram shows a molecule of tRNA.

Figure 2

The region labelled X is called

A   a codon.   B   an anticodon.
C   a regulator gene.   D   an amino acid attachment site.

6   The diagram below shows two chromosomes. The lettered regions represent genes.

chromosome 1   chromosome 2

| P | Q | R | S | T | U | V | W | | E | F | G | H |

Figure 3

Which of the following would result if a translocation occurred between chromosomes 1 and 2?

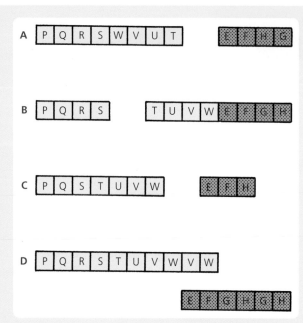

Figure 4

7   Bacterial species P in the diagram below is known to be resistant to the antibiotic streptomycin and able to transfer this characteristic to bacterial species R by conjugation. Which letter indicates a possible location of the gene for resistance to streptomycin?

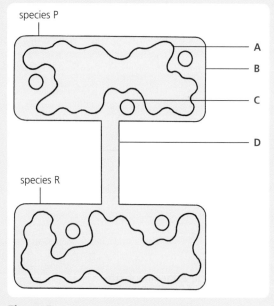

Figure 5

8   The following list gives descriptions of three aspects of the relationship between an enzyme and its substrate(s).

P = way in which molecules of two reactants are held together as determined by the enzyme's active site.

Q = state of close molecular contact resulting from a change in shape of the enzyme's active site to accommodate its substrate.

R = complementary relationship between a molecule of the enzyme and its substrate.

Which line in the table below correctly matches these aspects with their descriptions?

	aspect of enzyme–substrate relationship		
	induced fit	**specificity**	**orientation**
A	Q	P	R
B	P	R	Q
C	R	Q	P
D	Q	R	P

Table 4

9   Which of the following is LEAST likely to be found at region X of the mitochondrion shown in the following diagram?

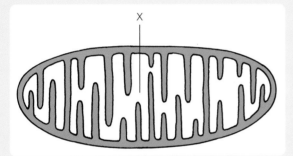

Figure 6

A   glucose   B   oxaloacetate   C   citrate   D   ADP

10 An amphibian's circulatory system is
A   complete and single.
B   incomplete and single.
C   complete and double.
D   incomplete and double.

11 Which of the following graphs best represents the variation in the metabolic rate of a naked man at rest in relation to the temperature of his environment?

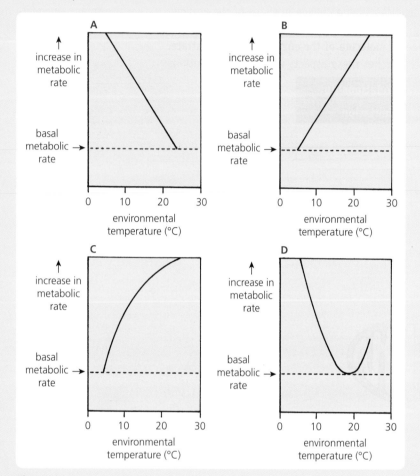

Figure 7

12 Some thermophiles live
   A in hot springs and obtain high energy electrons from inorganic molecules to make ADP.
   B submerged in ice and obtain high energy electrons from organic molecules to make ADP.
   C in hot springs and obtain high energy electrons from inorganic molecules to make ATP.
   D submerged in ice and obtain high energy electrons from organic molecules to make ATP.

13 The graph below indicates the results from closely monitoring the changes that take place during the fermentation of a closed batch of wine.

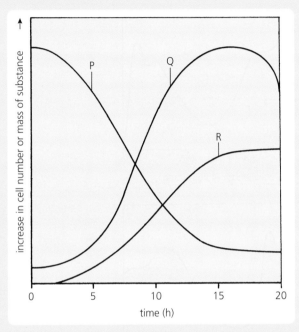

Figure 8

Which row in the following table correctly identifies P, Q, and R?

	P	Q	R
A	alcohol	yeast cells	glucose
B	glucose	alcohol	yeast cells
C	yeast cells	glucose	alcohol
D	glucose	yeast cells	alcohol

Table 5

14 Which row in the following table is NOT correct?

	DNA recombinant technology 'tool'	use of 'tool'
A	DNA ligase	to cleave open bacterial plasmids
B	bacterial plasmid	to transfer a DNA sequence from one genome to another
C	restriction endonuclease	to cut up DNA into fragments
D	artificial chromosome	to transfer a long length of DNA to a recipient cell

Table 6

15 Which of the following graphs represents the action spectrum of photosynthesis?

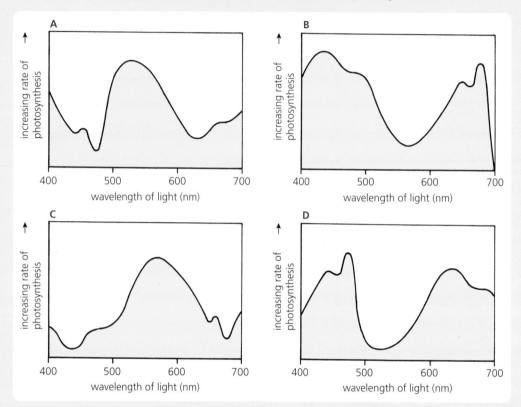

Figure 9

16 Burying perennial weeds by ploughing to a depth at which they die and decompose is an example of a control method that is

A preventative and effected by chemical means.

B traditional and effected by non-chemical means.

C curative and effected by non-chemical means.

D cultural and effected by chemical means.

17 Which of the following diagrams represents reciprocal altruism?

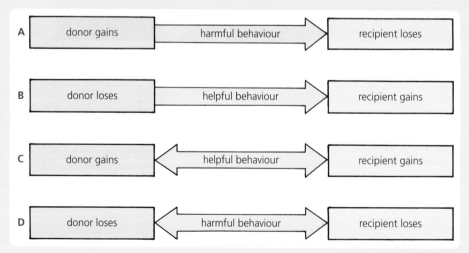

Figure 10

18 Which of the following are BOTH subordinate responses shown by a young wolf in the presence of a pack leader?
   A   head lowered and ears cocked
   B   ears flattened and eyes averted
   C   hackles raised and tail lowered
   D   eyes staring and teeth bared

19 Imagine that four different uninhabited volcanic islands are simultaneously formed off the coast of a mainland populated by a diverse community. After several years, species richness would be greatest on the island that was
   A   small and remote from the mainland.
   B   large and remote from the mainland.
   C   small and close to the mainland.
   D   large and close to the mainland.

20 The following graph charts the effect of increased intensity of fishing on four species of edible fish caught in the North Sea. (Sustained yield means the maximum yield of fish that can be maintained from year to year). Which species of fish is most likely to be the slowest-growing one?

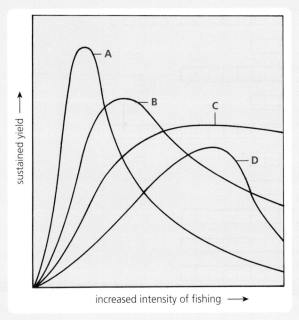

Figure 11

Answer grid

QUESTION NUMBER

| 1 | 2 | 3 | 4 | 5 | 6 | 7 | 8 | 9 | 10 | 11 | 12 | 13 | 14 | 15 | 16 | 17 | 18 | 19 | 20 | 21 | 22 | 23 | 24 | 25 |

TEST NUMBER

1

2

3

4

5 and 6

7

8

9.1

9.2

10

11

12

13

14

15 and 16

17

18

19

20

21

22

23

24

Exam 1

Exam 2

| 1 | 2 | 3 | 4 | 5 | 6 | 7 | 8 | 9 | 10 | 11 | 12 | 13 | 14 | 15 | 16 | 17 | 18 | 19 | 20 | 21 | 22 | 23 | 24 | 25 |

QUESTION NUMBER

	1	2	3	4	5	6	7	8	9	10	11	12	13	14	15	16	17	18	19	20	21	22	23	24	25
1	A	C	B	B	D	C	C	D	B	A															
2	D	B	A	B	C	A	C	B	C	D															
3	A	C	B	B	C	B	A	B	D	D	A	C	D	A	A	B	C	D	A	D	B	C	C	D	D
4	B	C	A	C	A	D	C	B	D	B	D	D	B	C	A										
5 and 6	A	C	A	D	A	D	B	C	C	B	D	C	A	B	B	B	C	A	B	C	C	D	D	A	D
7	C	C	D	A	D	C	A	B	A	B	B	C	C	D	B	D	C	A	D	D	B	A	A	A	B
8	D	A	C	C	B	D	A	D	A	C	B	D	B	B	C										
9.1	D	A	B	A	C	B	C	D	D	C	D	C	B	A	B	D	A	C	B	A					
9.2	A	A	C	D	D	A	B	C	C	D	A	D	B	C	B										
10	B	A	B	C	C	C	D	A	C	D	B	C	D	A	D	B	C	A	B	A	B	D	A	A	D
11	C	B	C	A	B	A	D	A	D	D	C	B	C	A	D										
12	B	D	B	B	A	D	A	D	A	C	C	C	D	B	C										
13	A	C	B	B	D	B	A	D	C	A	D	D	A	B	C	C	B	D	C	A					
14	A	C	D	D	A	C	D	B	C	D	A	A	C	B	C	B	D	A	B	B					
15 and 16	C	A	D	A	B	D	C	D	B	C	A	C	D	A	D	A	B	B	C	B					
17	A	C	B	B	C	D	B	A	D	C	B	D	A	A	A	D	B	D	D	A	B	C	C	C	D
18	D	C	A	A	B	B	C	C	B	B	D	A	D	A	C	A	D	D	C	B					
19	A	C	D	D	A	D	B	D	A	B	C	A	A	B	C	D	A	C	B	C	C	C	B	B	D
20	A	D	D	B	A	C	B	A	B	C	C	A	D	C	B										
21	B	D	C	D	A	C	C	A	B	D	C	C	B	D	A	A	C	D	B	D	A	B	D	B	A
22	C	B	D	A	D	B	D	C	C	A	B	A	D	B	A	B	C	C	D	A					
23	C	A	A	D	B	A	C	C	D	B															
24	C	C	D	B	C	D	A	D	C	B	B	A	B	D	A										
Exam 1	C	C	D	A	D	C	D	B	A	A	B	A	B	B	D	C	A	C	B	D					
Exam 2	C	A	C	B	D	B	C	D	A	D	A	C	D	A	B	B	C	B	D	A					

TEST NUMBER

	1	2	3	4	5	6	7	8	9	10	11	12	13	14	15	16	17	18	19	20	21	22	23	24	25